Komil Olimdzhanovich Mahmudov

Cauchy problems for the generalized system of Maxwell's equations

Komil Olimdzhanovich Mahmudov

Cauchy problems for the generalized system of Maxwell's equations

Monograph

ScienciaScripts

Imprint

Cover image: www.ingimage.com

This book is a translation from the original published under ISBN 978-620-7-47074-7.

Publisher:
Sciencia Scripts
is a trademark of
Dodo Books Indian Ocean Ltd. and OmniScriptum S.R.L publishing group

120 High Road, East Finchley, London, N2 9ED, United Kingdom
Str. Armeneasca 28/1, office 1, Chisinau MD-2012, Republic of Moldova, Europe
Printed at: see last page
ISBN: 978-620-8-28121-2

MINISTRY OF HIGHER EDUCATION, SCIENCE AND INNOVATIONS OF THE REPUBLIC OF UZBEKISTAN
SAMARKAND STATE UNIVERSITY

On the rights of the manuscript
UDC: 517.946

MAKHMUDOV KOMIL OLIMJANOVICH

CAUCHY PROBLEMS FOR THE GENERALIZED SYSTEM OF MAXWELL'S EQUATIONS

MONOGRAPHY.

2024

UDC 517.946

K.O.Makhmudov Cauchy problems for generalized system of Maxwell's equations (Monograph). Lambert Academic Publishing, **2024, 78 pp.**

For an elliptic complex of first order differential operators on a smooth manifold, we define a system of equations which can be viewed as abstract Maxwell equations. The formal theory of this system turns out to be very similar to the theory of classical Maxwell equations. We consider the Cauchy problem for the Maxwell equations in the theory of electrodynamics in a bounded region in Euclidean space. Both the tangent component and the normal component of the desired vector function are defined on a part of the boundary of the domain. We state a reasonable solvability condition and a Carleman formula for its solution. This problem is densely solvable and uncorrectable, making the standard calculus of Fourier integral operators inapplicable. If the surface where the data and the Cauchy data themselves are given is real-analytic, then we can apply the Cauchy-Kovalevsky theorem to guarantee the existence of a local solution. We use the special structure of Maxwell's equation to derive explicit global solvability conditions and construct an approximate solution

TABLE OF CONTENTS

INTRODUCTION

Numerous scientific and applied researches in the field of mathematics all over the world show that mathematical models of many processes lead to the study of incorrect boundary value problems for partial differential equations. The object of applied research on conditional correctness and construction of approximate solutions by set values on a part of the boundary of the domain, among partial differential equations of elliptic type, is especially important in hydrodynamics, geophysics, signal transmission theory, geological exploration and electrodynamics. The study of solutions of non-correct problems and the construction of a family of regularizing solutions was the impetus for the initiation of studies of the correctness class when continuing from the compacta. Due to the applied importance of the study, the Cauchy problem for a system of Maxwell-type equations in a spatial domain is an urgent problem of modern mathematical science.

At present, in the world of research in solving non-correct boundary value problems for the system of equations of Maxwell type elliptic by A.Douglas-L.Nirenberg and associated with the de Rama complex, the construction of regularized solution and obtaining the criterion of solvability plays a special role. In this targeted scientific study, the main objectives are to obtain an integral representation for solutions of the equation in a bounded domain, to construct the Carleman formula using the method of bases with double orthogonality, and to construct the Carleman function in explicit form, as well as to evaluate the conditional stability of solutions of the Cauchy problem and the criterion of solvability.

Scientists of our country pay more attention to actual directions of fundamental sciences with scientific and practical applications, in particular, they pay special attention to research of those problems of theory of differential equations and mathematical physics, which have practical application in fundamental sciences, and in particular, research of various boundary value problems for partial differential equations of elliptic type. The main tasks and directions of activity of mathematical science are to conduct research at the level of international standards in the priority directions of "Algebra and its applications, differential equations and their applications, mathematical modeling of nonlinear systems, dynamic systems and their applications, stochastic analysis, medical and biological informatics, and inquisitive mathematics "[1]. The development of the theory of partial differential equations and the theory of conditionally correct problems is of great importance in ensuring the fulfillment of this decree.

The subject and object of research of this dissertation is in line with the problems that are included in the subject of the tasks noted in the Decrees of the President of the Republic of Uzbekistan UP-4947 dated February 7, 2017 "On the strategy of action for further development of the Republic of Uzbekistan", № UP -2789 dated February 17, 2017 "On further improvement of the activities of the Academy of Sciences, organization, management and financing of research activities" and PP-4387 dated July 9, 2019 "On measures of state support for the further development of mathematics education and

[1] Decree of the President of the Republic of Uzbekistan PP-4387 of July 9, 2019 "On measures of state support for the further development of mathematics education and science, as well as indigenous s.Romanovsky Academy of Sciences of the Republic of Uzbekistan"

science, as well as on measures of state support for the further development of mathematics education and science, as well as on measures of state support for the further development of mathematics education and science".

The first results from the point of view of practical importance of non-correlated problems and narrowing the class of possible solutions to a compact, bringing the problems to stability were obtained in the work of A.N.Tikhonov. In the works of M.M.Lavrentiev, where estimates characterizing the stability of the spatial problem in the class of bounded solutions of the Cauchy problem for the Laplace equation and some other incorrect problems of mathematical physics in a straight cylinder, as well as for an arbitrary spatial domain with a sufficiently smooth boundary are obtained. S.N.Mergelyan for functions inside a sphere. E.M.Landis obtained estimates characterizing the stability of a spatial problem for an arbitrary elliptic equation; V.K.Ivanov obtained estimates in an infinite strip.

The Carleman function for the Laplace and Helmholtz equations is constructed by Sh.Yarmukhamedov when part of the boundary of the region is the surface of a cone, and by A.A.Shlapunov when part of the boundary is the surface of a sphere. The construction of the Carleman function by Sh.Y.Yarmukhamedov is based on the application of methods of the theory of functions. In this case, the fundamental solution is approximated on the conic part of the boundary of the domain. The construction by A.A.Shlapunov of the Carleman function is based on the approximation of the fundamental solution of the Laplace equation on the spherical part of the boundary of the domain by homogeneous harmonic polynomials, the construction by N.N.Tarkhanov and A.A.Shlapunov - for general elliptic systems in terms of bases with double orthogonality, by A.L.Buchheim and E.V.Arbuzov - for the Helmholtz equation and for the system of electrodynamic equations in the plane.

In the works of T.Ishankulov, E.Jabborov, O.Makhmudov and I.Niyozov - Cauchy problem for the system of Navier-Stokes equations and for the system of equations of elasticity theory in multidimensional space, E.N.Sattorov - Cauchy problem for the generalized Cauchy-Riemann system, for the homogeneous system of Maxwell equations is studied. In the work of K.O.Makhmudov, O.I.Makhmudov, N.N.Tarkhanov - for the system of Maxwell-type equations the Carleman formula is constructed . Incorrect boundary value problems for high order equations with self-adjoint operator type coefficients and equations of mixed-composite type have been the subject of research by Romanov V.G., S.I.Kabanikhin, K.S.Fayazov and their students. Inverse and non-correlated problems for elliptic and hyperbolic type equations were the subject of research by A.Khaidarov, A.Serikbaev and D.K.Durdiev, and integral geometry was studied by Akram Begmatov.

The following work has been done in the monograph:

The monograph is devoted to Cauchy problems for the system of Maxwell's equations. The system of Maxwell's equations is described in terms of the de Rham complex in R^3 i.e. the system of Maxwell's equations is represented in the language of differential forms [48],[51],[53].

The ellinticity of the system of Maxwell's equations according to A. Duglis and L. Nirenberg is considered. Here we give a construction of the matrix of the left fundamental solution of the system of electrodynamics equations of special form. The generalized Stratton-Chu formula obtained here is the main working tool of the present monograph. Solvability conditions are obtained and the Carleman formula is constructed using the method of bases with double orthogonality, i.e., we construct an explicit formula that recovers the solution of the system of Maxwell's equations in a bounded region , X in R^3 through the given tangential component and normal component of the solution on the part S of the boundary of the region.

In the final chapter we formulate Maxwell's equations in the context of arbitrary elliptic complexes on a compact manifold $X \subset R^n$ with edge. Here we generalize the Stratton-Chu formula to solutions of abstract Maxwell equations. In the monograph, not only the solvability conditions for the Cauchy problem are obtained, but also the Carleman formula for the system of Maxwell-type equations in . R^n

CHAPTER I. THE SYSTEM OF MAXWELL'S EQUATIONS FOR THE ELECTROMAGNETIC FIELD

§ 1.1 Ellipticity of the system of Maxwell's equations in the sense of A. Duglis - L. Nirenberg

A summary of Maxwell's equations.

Maxwell's equations contain all the basic laws of electric and magnetic fields, including electromagnetic induction, and are therefore the general equations of the electromagnetic field in resting media. Since Maxwell's equations form the basis of the theory of electromagnetism, let us consider briefly the physical meaning of these equations:

The conduction current flowing through the conductors and changes in the electric field in time lead to the appearance of a vortex magnetic field:

There is an eddy magnetic field around any current conductor and around any alternating electric field

$$rotH = j + \frac{\partial D}{\partial t} \quad (1.0.5)$$

Any change in the magnetic field in time causes a vortex electric field to appear:

At any change in the magnetic field, a vortex electric field appears, which is proportional to the rate of change of the magnetic field induction

$$rotE = -\frac{\partial B}{\partial t} \quad (1.0.6)$$

The system of Maxwell's equations explains the whole picture of electromagnetic phenomena from a single point of view. They are used to calculate fields according to given distributions in space of currents and charges.

This system of equations is supplemented by the so-called system of material equations, which characterize the individual properties of the material medium filling the space:

$$D = \varepsilon E; \quad B = \mu H; \quad j = \sigma E; \tag{1.0.7}$$

From Maxwell's equations follows the existence of electromagnetic waves, that is, such an electromagnetic field that is able to exist independently, in the absence of electric charges and currents.

The importance of Maxwell's equations in the creation of a unified theory of the electromagnetic field.

1. Maxwell's theory is a consistent theory of a single EMF field produced by an arbitrary system of charges and currents. This theory solves the main problem of electrodynamics - the characteristics of electric and magnetic fields are found for a given distribution of charges and currents. This theory was a generalization of the most important laws describing electric and magnetic phenomena (similar to Newton's equations and the beginnings of thermodynamics).

2. Maxwell's theory considers macroscopic fields, which are created by macrocharges and macrocurrents. Distances from the sources of fields to the points under consideration are much larger than the sizes of atoms. Periods of change of alternating electric and magnetic fields are much larger than periods of internal processes.

3. Maxwell's theory is phenomenological in nature. It does not consider the internal mechanism of phenomena in the medium. The medium is described by three quantities , ε μ and . σ

4. Maxwell's theory is a theory of close action, according to which electric and magnetic interactions occurring in electric and magnetic fields , propagate at a finite speed equal to the speed of light in a given medium.Then, substituting (1.0.7) into (1.0.5) and (1.0.6) respectively we obtain:

$$-\varepsilon(d/dt)E + rotH = \sigma E,$$
$$\mu(d/dt)H + rotE = 0.$$

see for interest [27], [26].

Although Maxwell's equations are applicable in all space and time, practical problems are finite, and solutions of Maxwell's equations inside the solution domain are coupled to the rest using boundary conditions, cf. [31],[33] and [22], etc., and started in time using initial conditions, cf. e.g. [23].

For simplicity, let us restrict ourselves to harmonic electromagnetic waves. Then electric and magnetic fields with electric permeability ε , magnetic permeability μ conductivity σ and frequency ω have asymptotic form

$$E(t,x) = (\varepsilon + i\sigma/\omega)^{-1/2} e^{-i\omega t} E(x),$$
$$H(t,x) = \mu^{-1/2} e^{-i\omega t} H(x).$$

From the system of Maxwell's equations with arbitrary time dependence

$$-\varepsilon(d/dt)E + rotH = \sigma E$$
$$\mu(d/dt)H + rotE = 0,$$

we have that the spatial coordinate-dependent parts of the fields E and H satisfy the system of Maxwell's equations in the harmonic regime

$$\begin{aligned} ikE + rotH = 0, \\ -ikH + rotE = 0, \end{aligned} \tag{1.1.1}$$

where the wave number k is given by the expression $k^2 = (\varepsilon + i\sigma/\omega)\mu\omega^2$. Let us choose the sign of k so that the condition $\mathrm{Im}k \geq 0$ is satisfied.

Let us write the system of Maxwell's equations (1.1.1) in scalar form in Cartesian coordinate system x_1, x_2, x_3 :

$$\begin{cases} -\frac{\partial E_2}{\partial x_3} + \frac{\partial E_3}{\partial x_2} - ikH_1 = 0, \\ \frac{\partial E_1}{\partial x_3} - \frac{\partial E_3}{\partial x_1} - ikH_2 = 0, \\ -\frac{\partial E_1}{\partial x_2} + \frac{\partial E_2}{\partial x_1} - ikH_3 = 0, \\ ikE_1 - \frac{\partial H_2}{\partial x_3} + \frac{\partial H_3}{\partial x_2} = 0, \\ ikE_2 + \frac{\partial H_1}{\partial x_3} - \frac{\partial H_3}{\partial x_1} = 0, \\ ikE_3 - \frac{\partial H_1}{\partial x_2} + \frac{\partial H_2}{\partial x_1} = 0. \end{cases} \tag{1.1.1'}$$

Then the system (1.1.1) can be written in the following form:

$$\begin{pmatrix} 0 & -\frac{\partial}{\partial x_3} & \frac{\partial}{\partial x_2} & -ik & 0 & 0 \\ \frac{\partial}{\partial x_3} & 0 & -\frac{\partial}{\partial x_1} & 0 & -ik & 0 \\ -\frac{\partial}{\partial x_2} & \frac{\partial}{\partial x_1} & 0 & 0 & 0 & -ik \\ ik & 0 & 0 & 0 & -\frac{\partial}{\partial x_3} & \frac{\partial}{\partial x_2} \\ 0 & ik & 0 & \frac{\partial}{\partial x_3} & 0 & -\frac{\partial}{\partial x_1} \\ 0 & 0 & ik & -\frac{\partial}{\partial x_2} & \frac{\partial}{\partial x_1} & 0 \end{pmatrix} \begin{pmatrix} H_1 \\ H_2 \\ H_3 \\ E_1 \\ E_2 \\ E_3 \end{pmatrix} = \begin{pmatrix} 0 \\ 0 \\ 0 \\ 0 \\ 0 \\ 0 \end{pmatrix} \qquad (1.1.1'')$$

Linear matrix differential operator $L(D)$ in $(1.1.1'')$, where

$$D = \left(\frac{\partial}{\partial x_1}, \frac{\partial}{\partial x_2}, \frac{\partial}{\partial x_3} \right) \qquad .$$

Consider the system of equations

$$L(x,D)U = g \qquad , (1.1.2)$$

here $g = g(x) = (g_1(x),...,g_N(x))^T$ is the given column vector,

$U = U(x) = (u_1(x),...,u_N(x))^T$ - unknown column vector, $,x = (x_1,...,x_u)$

N and n

- given natural numbers, the upper index T is the transpose sign. The symbol $L(x,D)$ denotes the operator defined by the formula

$$(L(x,D)U)_i = \sum_{j=1}^{N} l_{ij}(x,D)u_j(x), \; i = \overline{1,...,N}, \qquad (1.1.3)$$

where

$$l_{ij}(x,D) = \sum_{|\alpha| \le n_j} a_{ij}^{\alpha}(x) D^{\alpha} \qquad , i,j = \overline{1,...,N} \;, (1.1.4)$$

differential operators,

$$D = (D_1,...,D_n) = (\frac{\partial}{\partial x_1},...,\frac{\partial}{\partial x_n}), \; D^{\alpha} = \frac{\partial^{|\alpha|}}{\partial x_1^{\alpha_1} \cdots \partial x_n^{\alpha_n}} \qquad , \alpha = (\alpha_1,...,\alpha_n) \; -$$

vector from the space R^n , whose components are non-negative integers, $|\alpha| = \alpha_1 + ... + \alpha_n$, $a_{ij}^{\alpha}(x)$ -defined functions, $n_j = \max_{1 \le i \le N} \deg l_{ij}(x, D)$, $j = \overline{1,..., N}.$

The symbol deg denotes the order of the senior derivative included in the operator . $l_{ij}(x, D)$

Let $\widetilde{l}_{ij}(x, D)$ be the main part of the order n_j of each operator $l_{ij}(x, D)$, i.e.

$$\widetilde{l}_{ij}(x, D) = \begin{cases} 0, & \text{если} \quad \deg l_{ij}(x, D) < n_j \\ \sum_{|\alpha| = n_j} a_{ij}^{\alpha}(x) D^{\alpha} & \text{если} \quad \deg l_{ij}(x, D) = n_j. \end{cases}$$

Following [7], the main part (by Petrovsky) of the operator

$$L(x, D) = \begin{pmatrix} l_{11}(x, D) & ... & l_{1N}(x, D) \\ ... & ... & ... \\ l_{N1}(x, D) & ... & l_{NN}(x, D) \end{pmatrix} \tag{1.1.5}$$

call the operator

$$\widetilde{L}(x, D) = \begin{pmatrix} \widetilde{l}_{11}(x, D) & ... & \widetilde{l}_{1N}(x, D) \\ ... & ... & ... \\ \widetilde{l}_{N1}(x, D) & ... & \widetilde{l}_{NN}(x, D) \end{pmatrix} \tag{1.1.6}$$

Fix arbitrarily the vector $\xi = (\xi_{1,...,} \xi_n) \in R^n$, point $x \in X$ and associate with (1.1.6) the matrix

$$\widetilde{L}(x, \xi) = \begin{pmatrix} \widetilde{l}_{11}(x, \xi) & ... & \widetilde{l}_{1N}(x, \xi) \\ ... & ... & ... \\ \widetilde{l}_{N1}(x, \xi) & ... & \widetilde{l}_{NN}(x, \xi) \end{pmatrix} \tag{1.1.7}$$

where

$$\tilde{l}_{ij}(x,D)=\begin{cases}0, & если \quad |\alpha|<n_j\\ \sum\limits_{|\alpha|=n_j} a_{ij}^{\alpha}(x)\xi_1^{\alpha_1}\cdots\xi_n^{\alpha_n} & если \quad |\alpha|=n_j.\end{cases}$$

Definition 1.1[5, p 90]. *The system of equations (1.1.2) is called Petrovsky elliptic in the domain* X *, if for any* $x\in X$ *and for any* $\xi\in R^n\setminus\{0\}$ *the following condition is satisfied*

$$\det\tilde{L}(x,\xi)\neq 0. \qquad (1.1.8)$$

Let us show that, the system of equations $(1.1.1'')$ is not Petrovsky elliptic in the domain $.X\subset R^3$

For the system $(1.1.1'')$ the operator $L(x,D)=L(D)$ has the form

$$L(D)=\begin{pmatrix} 0 & -\frac{\partial}{\partial x_3} & \frac{\partial}{\partial x_2} & -ik & 0 & 0\\ \frac{\partial}{\partial x_3} & 0 & -\frac{\partial}{\partial x_1} & 0 & -ik & 0\\ -\frac{\partial}{\partial x_2} & \frac{\partial}{\partial x_1} & 0 & 0 & 0 & -ik\\ ik & 0 & 0 & 0 & -\frac{\partial}{\partial x_3} & \frac{\partial}{\partial x_2}\\ 0 & ik & 0 & \frac{\partial}{\partial x_3} & 0 & -\frac{\partial}{\partial x_1}\\ 0 & 0 & ik & -\frac{\partial}{\partial x_2} & \frac{\partial}{\partial x_1} & 0 \end{pmatrix}.$$

Highlighting the main part of the operator $L(D)$ by Petrovsky, we find $n_1=n_2=...=n_6=1$. Therefore, the main part of the operator $L(D)$ by Petrovsky is

$$\widetilde{L}(D)=\begin{pmatrix} 0 & -\frac{\partial}{\partial x_3} & \frac{\partial}{\partial x_2} & 0 & 0 & 0 \\ \frac{\partial}{\partial x_3} & 0 & -\frac{\partial}{\partial x_1} & 0 & 0 & 0 \\ -\frac{\partial}{\partial x_2} & \frac{\partial}{\partial x_1} & 0 & 0 & 0 & 0 \\ 0 & 0 & 0 & 0 & -\frac{\partial}{\partial x_3} & \frac{\partial}{\partial x_2} \\ 0 & 0 & 0 & \frac{\partial}{\partial x_3} & 0 & -\frac{\partial}{\partial x_1} \\ 0 & 0 & 0 & -\frac{\partial}{\partial x_2} & \frac{\partial}{\partial x_1} & 0 \end{pmatrix},$$

Write out the characteristic polynomial

$$\det \widetilde{L}(\xi)=\begin{vmatrix} 0 & -\xi_3 & \xi_2 & 0 & 0 & 0 \\ \xi_3 & 0 & -\xi_1 & 0 & 0 & 0 \\ -\xi_2 & \xi_1 & 0 & 0 & 0 & 0 \\ 0 & 0 & 0 & 0 & -\xi_3 & \xi_2 \\ 0 & 0 & 0 & \xi_3 & 0 & -\xi_1 \\ 0 & 0 & 0 & -\xi_2 & \xi_1 & 0 \end{vmatrix}.$$

The calculations give:

$$\det \widetilde{L}(\xi)=0$$

for any $\xi \in R^3$, i.e. the system$(1.1.1'')$ is not elliptic according to Petrovsky.

The given definition of the elliptic by Petrovsky system is simple enough, but turned out to be not broad enough.

The most general by now definition of an elliptic system belongs to A. Duglis and L. Nirenberg (note that there are also intermediate definitions, for example, strongly elliptic systems).

When determining the ellipticity of the system (1.1.2) by Duglis-Nirenberg, the principal part of the operator (1.1.5) is defined in a different way and it may include lower derivatives. Suppose that there exist two sets of integers

$$s=(s_1,...,s_N) \qquad \text{and}\, t=(t_1,...,t_N)\ . \quad (1.1.9)$$

satisfying the condition $\deg l_{ij}(x,D)\le s_i+t_j$ for all those $i,j=\overline{1,...,N}$, for which $l_{ij}(x,D)\neq 0$. Thus $l_{ij}(x,D)=0$, if $.\, s_i+t_j<0$

The main part of the Duglis-Nirenberg operator $L(x,D)$ corresponding to the sets $s=(s_1,...,s_N)$, $t=(t_1,...,t_N)$, let us call the matrix differential operator

$$\widetilde{L}_{s,t}(x,D)=\begin{pmatrix} \widetilde{l}_{11}^{s,t}(x,D) & ... & \widetilde{l}^{s,t}{}_{1N}(x,D) \\ ... & ... & ... \\ \widetilde{l}_{N1}^{s,t}(x,D) & ... & \widetilde{l}^{s,t}{}_{NN}(x,D) \end{pmatrix} \qquad (1.1.10)$$

where differential operators $\widetilde{l}_{ij}^{s,t}(x,D)$

$$\widetilde{l}_{ij}(x,D)=\begin{cases} 0, & \textit{если}\ \deg l_{ij}(x,D)<s_i+t_j\ \ \textit{или}\ \ l_{ij}(x,D)=0 \\ \sum\limits_{|\alpha|=n_j} a_{ij}^{\alpha}(x)D^{\alpha} & \textit{если}\ \ \deg l_{ij}(x,D)=s_i+t_j. \end{cases}$$

Definition 1.2[5, p 93]. *The system of equations (1.1.2) is called elliptic in the Douglas-Nirenberg region* X *, if for any* $x\in X$ *and for any* $\xi\in R^n\setminus\{0\}$ *the following condition holds*

$$\det \tilde{L}_{s,t}(x,\xi)\neq 0 \qquad .\ (1.1.11)$$

Let us show that, the classical system of Maxwell's equations (1.1.1) is Duglis-Nirenberg elliptic in the region $.\, X\subset R^3$

The fact that the system of Maxwell's equations is elliptic according to Duglis-Nirenberg follows from a well-known result of L.R. Volevich. Namely, from the fact that the determinant of the system of Maxwell's equations is an elliptic scalar operator [1], [2].

For the system$(1.1.1'')$ we choose sets (1.1.9), e.g., of the form $s=(0,0,0,0,0,0)$ and $t=(0,0,0,0,0,0)$. Corresponding to (1.1.10) the main part of the operator

$$L(D)=\begin{pmatrix} -ik & 0 & 0 & 0 & -\frac{\partial}{\partial x_3} & \frac{\partial}{\partial x_2} \\ 0 & -ik & 0 & \frac{\partial}{\partial x_3} & 0 & -\frac{\partial}{\partial x_1} \\ 0 & 0 & -ik & -\frac{\partial}{\partial x_2} & \frac{\partial}{\partial x_1} & 0 \\ 0 & -\frac{\partial}{\partial x_3} & \frac{\partial}{\partial x_2} & ik & 0 & 0 \\ \frac{\partial}{\partial x_3} & 0 & -\frac{\partial}{\partial x_1} & 0 & ik & 0 \\ -\frac{\partial}{\partial x_2} & \frac{\partial}{\partial x_1} & 0 & 0 & 0 & ik \end{pmatrix}$$

looks like

$$\widetilde{L}_{s,t}(D)=\begin{pmatrix} -ik & 0 & 0 & 0 & 0 & 0 \\ 0 & -ik & 0 & 0 & 0 & 0 \\ 0 & 0 & -ik & 0 & 0 & 0 \\ 0 & 0 & 0 & ik & 0 & 0 \\ 0 & 0 & 0 & 0 & ik & 0 \\ 0 & 0 & 0 & 0 & 0 & ik \end{pmatrix},$$

and its determinant

$$\det.\widetilde{L}_{s,t}(D)=k^6\neq 0$$

Using Definition 1.2, we conclude that the system (1.1.1) is Douglas-Nirenberg elliptic.

§ 1.2. Criterion of solvability of the Cauchy problem for Helmholtz equations

Let X be a bounded region in R^3 , and S – a smooth closed surface in X , partitioning it into two connected components X^+ and X^- oriented as a boundary . X^-

Let the function . $u(x) \in C^2(X^-) \cap C^1(\overline{X^-})$ Consider the Cauchy problem for the Helmholtz equation in the X^-

$$(\Delta + k^2)u(x) = 0 \qquad , x \in X^- ,$$

$$u(x)|_S = u_0(y); \ \frac{\partial u(x)}{\partial n}|_S = u_1(y) \quad , \tag{1.2.1}$$

where k^2 is a real number, Δ – the Laplace operator.

This problem is incorrect and it has at most one solution, i.e.: 1) the solution does not exist for any data; 2) the solution does not depend continuously on the Cauchy data at S (see 1.1 below). In A.A.Shlapunov [15 , 16] a criterion of solvability of the Cauchy problem for elliptic equations and systems is given. In the present paper we consider an analogous problem for the Helmholtz equation .

The instability of the solution of the problem (1.2.1) follows from the following example.

1.1-Example **.** Let S be a "piece" of the plane $\{x_3 = 0\}$ in R^3 and

$$u(x) = \sin mx_1 \cdot \sin mx_2 \cdot sh\ x_3 \sqrt{2m^2 - k^2},\ m > k .$$

It is easy to verify that the function $u(x)$ satisfies the Helmholtz equation in R^3 . Indeed,

$$(\Delta + k^2)u(x) = \Delta(\sin mx_1 \cdot \sin mx_2 \cdot sh x_3 \sqrt{2m^2 - k^2}) + k^2 u(x) =$$
$$= (-m^2 - m^2 + 2m^2 - k^2)u(x) + k^2 u(x) = 0.$$

In addition,

$$u(x_1, x_2, 0) = 0, \frac{\partial u(x_1, x_2, 0)}{\partial x_3} = \sqrt{2m^2 - k^2} \cdot \sin mx_1 \cdot \sin mx_2 .$$

However, at each point $x = (x_1, x_2, x_3)$ $x_1 \neq 0,$ $x_2 \neq 0,$ and $x_3 > 0$ we have:

$$\lim_{m \to \infty} u(x) = \infty .$$

Let us denote by $\Phi(x, y)$ the fundamental solution of the Helmholtz equation in the R^3

$$\Phi(x, y) = -\frac{1}{4\pi} \cdot \frac{\exp(ik|x-y|)}{|x-y|} , x \neq y .$$

In the proposition that u_0 и u_1 are summable functions on S, let us assume

$$\mathrm{F}(x) = \int_S \left(u_0(y) \frac{\partial \Phi(x, y)}{\partial n_y} - u_1(y) \Phi(x, y) \right) dS_y \; (x \in X \setminus S) .$$

Clearly, F satisfies the Helmholtz equation everywhere outside S , and let $F^{\mp}$ imply a contraction of F by . $X^{\mp}$

Lemma 1.1. *Let $S \in C^2$, a $u_0 \in C^1(S)$ and $u_1 \in C(S)$ be summable functions on S , then the function F^+ continues smoothly to $X^+ \cup S$ if and only if F^- continues smoothly to . $X^- \cup S$.*

Proof. Let us use the fact of existence of a smooth function $\hat{u}$, given in some neighborhood S of X such that , $\hat{u}|_s = u_0$ $\frac{\partial \hat{u}}{\partial n}|_s = u_1$ ([8] lemma 29.5).

If , $x^0 \in S$ $n(x^0) -$ is the unit vector normal to S at the point x^0 and $|\alpha| \leq 1,$ then

$$\lim_{\varepsilon \to 0} \left(\partial^\alpha F(x^0 - \varepsilon n(x^0)) - \partial^\alpha F(x^0 + \varepsilon n(x^0)) \right) = \partial^\alpha \hat{u} , \qquad (1.2.2)$$

and the limit is reached uniformly on compact subsets of $.S$

Suppose, for example, F continues smoothly into $X^{-} \cup S$. Fix a multi-index $|\alpha| \leq 1$, then

$$\lim_{\varepsilon \to 0} \partial^{\alpha} F^{+}(x^{0} + \varepsilon n(x^{0})) = \partial^{\alpha} F^{-}(x^{0}) - \partial^{\alpha} \hat{u}(x^{0}).$$

Let's pre-define $\partial^{\alpha} F^{+}$ to $X^{+} \cup S$ as follows:

$$\partial^{\alpha} F^{+}(x) = \begin{cases} \partial^{\alpha} F^{+}(x), & x \in X^{+}, \\ \partial^{\alpha} F^{-}(x) - \partial^{\alpha} \hat{u}(x), & x \in S. \end{cases}$$

We show that $\partial^{\alpha} F^{+}$ is continuous in $X^{+} \cup S$. Let's fix an arbitrary point $x^{0} \in S$ and $E > 0$. Since $\partial^{\alpha} F^{+}$ is continuous on S, , there is $\delta_{0} > 0$ such that if $x^{1} \in S$ and $|x^{1} - x^{0}| < \delta_{0}$, then $|\partial^{\alpha} F^{+}(x^{1}) - \partial^{\alpha} F^{+}(x^{0})| < E/2$. Reducing δ_{0}, if necessary, we can assume that $K = \overline{B(x^{0}, \delta_{0})} \cap S -$ is a compact subset of S.

Since the surface S is smooth, we can choose a number $0 < \delta < \delta_{0}/2$ such that each point $x \in X^{+} \cap B(x^{0}, \delta)$ is represented as $x = x^{1} + \varepsilon n(x^{1})$, where $x^{1} \in S$, and $\varepsilon = dist(x, S)$. Then $\varepsilon > \delta$, so $|x^{0} - x| \leq |x^{1} - x| + |x^{0} - x^{1}| < \delta_{0}$, i.e., $.x^{1} \in K$

Given that the limit in (2.2.2) is uniformly attained on compact subsets of S, and reducing, if necessary, δ , we can achieve then that at $x^{1} \in K$ and $0 < \varepsilon < \delta$ the inequality is satisfied.

$$|\partial^{\alpha} F^{+}(x^{1} + \varepsilon n(x^{1})) - \partial^{\alpha} F^{+}(x^{1})| < \varepsilon/2.$$

Let now $x \in X^{+} \cap B(x^{0}, \delta)$, then for some $x^{1} \in K$ and $0 < \varepsilon < \delta$ we have $x = x^{1} + \varepsilon n(x^{1})$. Since

$$|\partial^{\alpha} F^{+}(x^{0}) - \partial^{\alpha} F^{+}(x)| \leq |\partial^{\alpha} F^{+}(x^{0}) - |\partial^{\alpha} F^{+}(x^{1})| + |\partial^{\alpha} F^{+}(x^{1} + \varepsilon n(x^{1})) - \partial^{\alpha} F^{+}(x^{1})| < \varepsilon$$

,

hence it follows that F^+ smoothly continues to $X^+ \cup S$, if F^- smoothly continues to $X^- \cup S$ and vice versa. The lemma is proved.

Theorem 1.2. *Let $S \in C^2$, and $u_0 \in C^1(S)$ and $u_1 \in C(S)$ be summable functions on S . Then for solvability of problem (1.2.1) it is necessary and sufficient that F^+ continues as a solution of the Helmholtz equation from X^+ to the region . X*

Necessity. Suppose that there exists a function u , which is a solution to problem (1.2.1). Consider the following function in X :

$$V(x) = \begin{cases} F^+(x), & x \in X^+ \\ F^-(x) - u(x), & x \in X^- \end{cases} \tag{1.2.3}$$

Note that for any subarea $S_1 \subset S$ there is some region $X_1 \subset X$ in X^- with piecewise smooth boundary such that $S_1 \subset \partial X_1$. It is clear that $u \in C^1(\overline{X_1})$ solutions of the Helmholtz equation in X_1 , therefore by Green's formula

$$u(x) = \int_{\partial X_1} \left(u(y) \frac{\partial \Phi(x,y)}{\partial n_y} - \frac{\partial u}{\partial n}(y) \Phi(x,y) \right) dS(y),\ x \in X_1 \quad . \tag{1.2.4}$$

Hence we have:

$$V(x) = F^-(x) - u(x) = \int_{S \setminus S_1} \left(u_0(y) \frac{\partial \Phi(x,y)}{\partial n_y} - u_1(y) \Phi(x,y) \right) dS(y) -$$

$$- \int_{\partial X_1 \setminus S_1} \left(u(y) \frac{\partial \Phi(x,y)}{\partial n_y} - \frac{\partial u}{\partial n}(y) \Phi(x,y) \right) dS(y). \tag{1.2.5}$$

The terms in the right-hand side of (1.2.5) are solutions of the Helmholtz equation in the neighborhood of S_1 and hence, since S_1 is arbitrary, F^- smoothly continues in the $X^- \cup S$.

Further, it follows from Lemma 1.1 that also F^+ smoothly continues to $X^+ \cup S$. Therefore, the contraction of $V^\mp$ by $X^\mp$ of the function V

smoothly continues to $X^{\mp} \cup S$. Moreover, from (1.2.2), for $x^0 \in S$ we have [8]

$$\lim_{\varepsilon \to 0} \left(V^-(x^0 - \varepsilon n(x^0)) - V^+(x^0 + \varepsilon n(x^0)) \right) = u_0,$$

$$\lim_{\varepsilon \to 0} \left(\frac{\partial V^-}{\partial n}(x^0 - \varepsilon n(x^0)) - \frac{\partial V^+}{\partial n}(x^0 + \varepsilon n(x^0)) \right) = u_1.$$

Thus, we conclude that the function V can be smoothly continued to the entire region X , by putting, for example, $V = F^- - u$ on the S.
It follows from Theorem 1.2 [8] that the function V , is smooth in X and satisfies the Helmholtz equation. It follows from (1.2.3) that V is the required continuation of F^+ in . X

Sufficiency. Let F^+ continue to X as solutions of the Helmholtz equation, denote the continued solution by V . Then it follows from the lemma that F^- smoothly continues into $X^- \cup S$. Let $u(x) = F^-(x) - V(x)$ $(x \in X^-)$.

Using the formulas (1.2.2) as in the proof of the necessity of Theorem 1.2.3, we will see that the contractions on S of the function u and its normal derivative $\frac{\partial u}{\partial n}$ are u_0 and u_1 respectively.

§ 1.3. Differential forms. Cohomologies of de Rham manifolds

A differential form w of degree p is a differential form of degree $p+1$, which is denoted by dw , and if w in local coordinates x is written as

$$w = \sum_{|I|=p} a_I(x) dx_I ,$$

then

$$dw = \sum_{|I|=p} da_I(x) \wedge dx_I \ .$$

Let us describe in more detail dw, given that $da_I(x)$ is the ordinary differential of the function $a_I(x)$:

$$dw = \sum_{|I|=p} da_I(x) \wedge dx_I = \sum_{|I|=p} \left(\frac{\partial a_I}{dx_1} dx_1 + \cdots + \frac{\partial a_I}{dx_n} dx_n \right) \wedge dx_I =$$
$$= \sum_{|I|=p} \sum_{|j|=1} \frac{\partial a_I}{\partial x_j} \wedge dx_j \wedge dx_{i_1} \wedge \cdots \wedge dx_{i_p} .$$

Properties of differential dw in a fixed local coordinate system $x = (x_1, ..., x_n)$.

1. $d(\varphi + \psi) = d\varphi + d\psi$, where $,\varphi \ \psi \ dw \in C^1$
2. Leibniz's generalized rule: $d(\varphi \wedge \psi) = d\varphi \wedge \psi + (-1)^{\deg \varphi} \varphi \wedge d\psi$
3. $d^2 w = d(dw) = 0$ where $w \in C^2$.

Note that the differential 0-form is a function.

There is a differentiation operator (external differential)

$$d : \Omega^p(R^n) \to \Omega^{p+1}(R^n),$$

is defined in such a way that it satisfies the following properties:

The outer differentiation operator raises the degree of the form by one and acts as follows:

$$\text{if } h \in \Omega^0(R^n),\ dh = \sum \frac{\partial h}{\partial x^l} dx^l ,$$

$$\text{if} \omega = \sum h_I dx_I, \text{ then } d\omega = \sum dh_I dx_I \ .$$

Example 1.2: If $\omega = xdy - ydx$, then $d\omega = dx \wedge dy - dy \wedge dx = 2dx \wedge dy$

Example 1.3. If $\omega = xdx + ydy + zdz.$

Then.

$$d\omega = d[xdx + ydy + zdz] =$$
$$= d(xdx) + d(ydy) + d(zdz) =$$
$$dx \wedge dx + dy \wedge dy + dz \wedge dz = 0$$

Note that the equality $d\omega = 0$ can be proved in another way. The point is that

$$\omega = df \text{ , where. } \frac{1}{2} d(x^2 + y^2 + z^2)$$

Then.

$$d\omega = d(df) = 0 \, .$$

Example1.4. Cartan distribution

$$\omega = dy - zdx \, .$$

Тогда

$$d\omega = d(dy - zdx) =$$
$$= d(dy) - d(zdx) = -dz \wedge dx = dx \wedge dz. \, .$$

The operator d , called outer differentiation, is a highly abstract generalization of the usual gradient, rotor and divergence considered in vector calculus by R^3 , as the following example partially illustrates.

It is not difficult to see that this property uniquely determines the outer derivative: it suffices for this fact that any 1-form is a locally finite linear combination of forms of the form dh (with coefficients from . $\Omega^0(R^3)$

Let's look at some special cases that are important from the point of view of history and applications. Let $M = R^3$. The mapping $d : \Omega^0(R^3) \to \Omega^1(R^3)$ is called a gradient in the application literature, the mapping $d : \Omega^1(R^3) \to \Omega^2(R^3)$ a rotor, and the mapping $d : \Omega^2(R^3) \to \Omega^3(R^3)$ a divergence.

For the case of the space R^3 the summands $\Omega^0(R^3)$ and $\Omega^3(R^3)$ Both one-dimensional, $\Omega^1(R^3)$ and $\Omega^3(R^3)$ are three-dimensional over C^∞ – smooth functions, so the following identifications are possible:

$$\{C^\infty - \text{гладкие функции}\} \cong \{0-\text{формы}\} \cong \{3-\text{формы}\},$$

$h \leftrightarrow \qquad h \qquad \leftrightarrow h\,dxdydz,$

u

$$\{C^\infty - \textit{гладкие векторные поля}\} \cong \{1-\text{формы}\} \cong \{2-\text{формы}\},$$

$$W = (h_1, h_2, h_3) \leftrightarrow h_1dx + h_2dy + h_3dz \leftrightarrow$$

$h_1dydz - h_2dxdz + h_3dxdy$.

On functions we have

$$dh = \frac{\partial h}{\partial x}dx + \frac{\partial h}{\partial y}dy + \frac{\partial h}{\partial z}dz$$

On 1-forms.

$$d(h_1dx + h_2dy + h_3dz) =$$
$$\left(\frac{\partial h_3}{\partial y} - \frac{\partial h_2}{\partial z}\right)dydz - \left(\frac{\partial h_1}{\partial z} - \frac{\partial h_3}{\partial x}\right)dxdz + \left(\frac{\partial h_2}{\partial x} - \frac{\partial h_1}{\partial y}\right)dxdy.$$

On the 2-form.x

$$d(h_1dydz - h_2dxdz + h_3dxdy) = \left(\frac{\partial h_1}{\partial x} + \frac{\partial h_2}{\partial y} + \frac{\partial h_3}{\partial z}\right)dxdydz.$$

Sequence

$$0 \longrightarrow \Omega^0 \xrightarrow{d^0} \Omega^1 \xrightarrow{d^1} \cdots \xrightarrow{d^{m-1}} \Omega^m \xrightarrow{d^m} \Omega^{m+1} \longrightarrow \cdots$$

groups (or, in particular, linear spaces) and their homomorphisms is called a cochain complex if $d^m \circ d^{m-1} = 0$ for any $m \geq 1$.

Usually instead of d^m you just write $.d$

The kernel of $Ker d^m$ homomorphisms of d^m is denoted by $Z^m(\Omega)$, and its elements are called cocycles of degree m. The image of $\operatorname{Im} d^{m-1}$ homomorphisms of d^{m-1} is denoted by $B^m(\Omega)$, and its elements are called cocycles of degree . m ,

De Rham's cohomology is a theory of cohomologies based on differential forms, and applied in theories of smooth and algebraic manifolds.Named after the Swiss mathematician de Rham. The k-dimensional cohomology group of de Rham manifolds M is usually denoted by $H^k_{dR}(M)$, [20, p. 22].

CHAPTER II. CAUCHY PROBLEMS FOR THE SYSTEM OF MAXWELL'S EQUATIONS

It is well known that the Cauchy problem for elliptic equations and systems is incorrect in all standard function spaces. Nevertheless, it naturally arises in applications: in hydrodynamics (as the Cauchy problem for holomorphic functions), in geophysics (as the Cauchy problem for the Laplace operator), in elasticity theory (as the Cauchy problem for the Lamé system), in electrodynamics (as the Cauchy problem for the Maxwell system), etc. A huge number of works by such famous mathematicians as J. Adamar, T. Karleman, Goluzin-Krylov, M.M. Lavrentiev, V. Ivanov, Sh. Yarmukhamedov and others are devoted to these problems. One of the first formulas restoring a holomorphic function in a region of one special kind by its values on a part of the boundary was proposed by Karleman, and formulas of this kind became known as Karleman formulas. All these formulas and many others, as well as their applications, are presented in Eisenberg's monograph. At the end of the 20th century, it was realized that the Cauchy problem for elliptic systems of linear equations is equivalent to another incorrect problem: the "analytic continuation" from a smaller open set to a larger one. For overdetermined elliptic systems, this approach, based on the method of integral representations and developed with the participation of L. Eisenberg, N. Tarkhanov, A. Shlapunov and others, also turned out to be quite productive theoretically and practically: simple formulas for exact and approximate solutions of the problem were constructed.

Proceeding from this, in this chapter the Cauchy problems for the system of Maxwell's equations with two methods are studied, i.e. the continuation formula for the stationary system of Maxwell's equations in the cap-type region as well as for the part of the ball is obtained.

Homogeneous polynomials

A homogeneous polynomial of degree ν in space R^n is **called-** polynomial of the form

$$P(x)=\sum_{|\alpha|=\nu} c_\alpha x^\alpha$$

where $\alpha=(\alpha_1,...,\alpha_n)$ is the multi-index, $\alpha_j \in Z_+ \ j=1,...n$, $|\alpha|=\alpha_1+...+\alpha_n$

$\alpha^\alpha=\alpha_1^{\alpha_1},...,\alpha_n^{\alpha_n})$, $c_\alpha \in R$. Let us denote by L_ν the set of all homogeneous polynomials of degree ν . Obviously, L_ν is a linear space. Suppose

$$D^\alpha=\frac{\partial^{|\alpha|}}{\partial x_1^{\alpha_1}...\partial x_n^{\alpha_n}}$$

If $P\in L_\nu$, then

$$P(D)=\sum_{|\alpha|=\nu} c_\alpha D^\alpha$$

- differential operator. In particular, if

$$P(x)=|x|^2=|x_1|^2+...+|x_n|^2,$$

then $P(D)=\Delta$ is the Laplace operator.

Let us introduce in the space L_ν the scalar product by putting

$$\langle P,Q\rangle=P(D)Q,\ P,Q\in L_\nu \qquad .$$

Let us check that the introduced operation is indeed scalar-
as a work of art. First of all, we note that

$$D^{\alpha}x^{\beta}=\begin{cases}0, & \alpha\neq\beta\\ \alpha!=\alpha_1!...\alpha_n!, & \alpha=\beta.\end{cases}$$

Thus $\langle P,Q\rangle$ takes scalar values. Linearity

of the function $\langle P,Q\rangle$ is obvious, and from the equality

$$D^{\alpha}x^{\beta}=D^{\beta}x^{\alpha}$$

commutativity follows. Since

$$\langle P,P\rangle=\sum_{|\alpha|=k}|c_{\alpha}|^2\alpha!,$$

then $\langle P,P\rangle=0$ if and only if . $P=0$

In the space L_{ν} the uninomials , x^{α} $|\alpha|=\nu$, form an orthogonal system

and hence are an orthogonal basis

of this space. The number of such one-members n_{ν} is equal to the

number of times-

personal multi-indexes $\alpha=(\alpha_1,...,\alpha_n)$ such that . $\alpha_1+...+\alpha_n=\nu$

To find this number, consider a $n-\nu-1$ urn, pick on-

luck $n-1$ urn, and put one ball in each of the other urns. It's not hard

make sure that the number n_{ν} equals the number of ways to select $n-1$

urns.

from $\nu+n-1$ urns, ie.

$$n_{\nu}=C_{n+\nu-1}^{n-1}=\frac{(n+\nu-1)!}{(n-1)!\nu!}.$$

Homogeneous harmonic polynomials

Homogeneous harmonic polynomials of degree ν are polynomials $P\in L_{\nu}$ such that $\Delta P=0$. We denote the set of homogeneous harmonic polynomials of degree ν by A_{ν} . It is clear that . $A_{\nu}\subset L_{\nu}$

Theorem 0.1 Every polynomial $P \in L_\nu$ can be represented as

$$P(x) = P_0(x) + |x|^2 P_1(x) + ... + |x|^{2l} P_l(x), \quad (2.0.1)$$

where $P_j(x) \in A_{\nu-2j}, \quad j = 0,...,l.$

Proof. Let us assume that $\nu \geq 2$ (at $\nu = 0{,}1$ $L_\nu = A_\nu$).

Consider the linear subspace

$$|x|^2 L_{\nu-2} = \{P \in L_\nu : P(x) = |x|^2 Q(x), \ Q \in L_{\nu-2}\}$$

Let us prove that its orthogonal complement is A_ν . Let $P_1 \in L_\nu$ and for all $P_1 \in L_{\nu-2}$

$$\langle R, P_1 \rangle = 0, \quad (2.0.2)$$

where $R(x) = |x|^2 Q(x).$ But

$$\langle R, P_1 \rangle = \Delta Q(D) P_1 = Q(D) \Delta P_1 = \langle Q, \Delta P_1 \rangle. \quad (2.0.3)$$

Let us choose $Q \in L_{\nu-2}$ such that $Q = \Delta L_{\nu-2}$. It follows from (2.0.3) that

$$\langle \Delta P_1, \Delta P_1 \rangle = 0..$$

Hence, $\Delta P_1 = 0$ and $\Delta P_1 \in A_\nu$. It also follows from (2.0.3) that the converse is also true, i.e. if $\Delta P_1 = 0$, then for all $Q \in \Delta L_{\nu-2}$ it follows that equality (2.0.2). Thus, it is proved that

$$L_\nu = A_\nu \oplus |x|^\nu L_{\nu-2}. \quad (2.0.4)$$

Let $P \in L_\nu$. Then there are such homogeneous polynomials $P_0 \in A_\nu$ and $Q_1 \in L_{\nu-2}$ that

$$P(x) = P_0(x) + |x|^2 Q_1(x).$$

If $k \geq 4,$, it can be similarly represented as follows $Q_1(x)$

$$Q_1(x) = P_1(x) + |x|^2 Q_2(x), \text{ (x)}, \ P_1 \in A_{k-2}, \ Q_1 \in L_{k-4}.$$

Thereby

$$P(x) = P_0(x) + |x|^2 P_1(x) + |x|^4 Q_1(x),$$

Continuing this process, we arrive at the representation (2.0.1).

Corollary 2.1. For **all** $k \geq 2$

$$a_k = \dim A_k = n_k - n_{k-2} = (n + 2k - 2)\frac{(n+k-3)}{(n-2)!k!}. \qquad (2.0.5)$$

Proof. From (2.0.5) we have

$$n_k = \dim L_k = \dim A_k + \dim |x|^k L_{k-2} = \dim A_k + n_{k-2}.$$

Hence (2.0.5) follows.

Due to the fact that $A_k = L_k$ at $k = 0,1,$ we have

$$a_0 = \dim A_0 = 1, \quad a_1 = \dim A_1 = n.$$

§2.1 The Stratton-Chu formula

We begin our consideration with the derivation of the Stratton-Chu formulae, which show that any solution of the system of Maxwell's equations (1.1.1) can be represented as an electromagnetic field excited by a combination of electric and magnetic dipoles located on a surface.

Theorem 2.1.*[21] Let $E, H \in C^1(X) \cap C(\overline{X})$ represent a solution to the system of Maxwell's equations (1.1.1) in the X*

Then the classical Stratton-Chu formulas take place

$$rot \int_{\partial X} [\nu(y), E(y)] \Phi(x,y) ds(y) -$$
$$- \frac{1}{ik} rotrot \int_{\partial X} [\nu(y), H(y)] \Phi(x,y) ds(y) =$$
$$= \begin{cases} -E(x), & x \in X, \\ 0, & x \in R^3 \setminus \overline{X}, \end{cases}$$

и

$$rot\int_{\partial X}[\nu(y),H(y)]\Phi(x,y)ds(y)-$$

$$-\frac{1}{ik}rotrot\int_{\partial X}[\nu(y),E(y)]\Phi(x,y)ds(y)=$$

$$=\begin{cases}-H(x), & x\in X,\\ 0, & x\in R^3\setminus\overline{X}.\end{cases}$$

Proof. We use the notations introduced in the proof of Green's formula for the Helmholtz equation and choose an arbitrary fixed point $x\in X$ and an arbitrary fixed unit vector $e\in R^3$. Using Maxwell's equations for E and H and the relation $rotrotrot\ e\Phi=k^2 rot\ e\Phi,$ we find

$$div\left\{[E,rot\,e\Phi]-\frac{1}{ik}[H,rotrot\,e\Phi]\right\}=0\ в\quad X\setminus\{x\}.$$

Hence, from Gauss's theorem we obtain

$$\int_{\partial X+B_{x,r}}\{(\nu(y),E(y),rot_y e\Phi(x,y))-$$

$$-\frac{1}{ik}(\nu(y),H(y),rot_y rot_y e\Phi(x,y))\}ds(y)=0.$$

Using Stokes' theorem and Maxwell's second equation, we have

$$\int_{B_{x,r}}(\nu(y),H(y),grad_y div_y e\Phi(x,y))=$$

$$-ik\int_{B_{x,r}}(\nu(y),E(y)),div_y e\Phi(x,y)ds(y).$$

Then, since at $B_{x,r}$ there is $\Phi(x,y)=O(1/r)$

и

$$div\,e\Phi(x,y)=\frac{(\nu(y),e)}{4\pi r^2}+O\left(\frac{1}{r^3}\right),$$

$$rot\,e\Phi(x,y)=\frac{[\nu(y),e]}{4\pi r^2}+O\left(\frac{1}{r^3}\right),$$

use the relation $rotrot\,e\Phi=k^2e\Phi+graddiv\,e\Phi$
и by direct calculation we obtain

$$\lim_{r\to\infty}\int_{B_{x,r}}\{(\nu(y),E(y),rot_y e\Phi(x,y))-$$

$$-\frac{1}{ik}(\nu(y),H(y),rot_y rot_y e\Phi(x,y))\}ds(y)=(e,E(x)).$$

Finally, due to the symmetry of the relation $grad_x\,\Phi(x,y)=-grad_y\Phi(x,y)$ и easy to show что

$$(\nu(y),E(y),rot_y e\Phi(x,y))=(e,E(y),rot_x[\nu(y),E(y)]\Phi(x,y))$$

и

$$(\nu(y),H(y),rot_y rot_y e\Phi(x,y))=(e,rot_x,rot_x[\nu(y),H(y)]\Phi(x,y))$$

Hence, we obtain that

$$\left(e,\int_{\partial X}\{rot_x[\nu(y),E(y)]\Phi(x,y)-\frac{1}{ik}rot_x rot_x[\nu(y),E(y)]\Phi(x,y)\}ds(y)+E(x)\right)=0.$$

Due to the arbitrariness of e , the validity of Theorem 2.1 for $x\in X.$

If $x\in R^3\setminus\overline{X},$ then the proof is obtained analogously from the identity

$$\int_{\partial X}\{(\nu(y),E(y),rot_y e\Phi(x,y))-\frac{1}{ik}(\nu(y),H(y),rot_y rot_y\Phi(x,y))\}ds(y)=0.$$

Now it is easy to obtain the representation for the field $H,$ by using the relation $H=(1/ik)rotE.$

Theorem 2.2[21] *The Cartesian components of any continuously differentiable solution of a system of Maxwell's equations are analytic functions of coordinates.*

In particular, the Cartesian components of solutions of the system of Maxwell's equations are twice continuously differentiable. Hence, we can use the vector identity

$$rotrotE=-\Delta E+graddivE,$$

Expressing the system of Maxwell's equations (1.1.1) in terms of differential forms leads to further notation and conceptual simplification. Let us introduce the de Rham complex in R^3

$$0 \longrightarrow \Omega^0(R^3) \xrightarrow{d^0} \Omega^1(R^3) \xrightarrow{d^1} \Omega^2(R^3) \xrightarrow{d^2} \Omega^3(R^3) \longrightarrow 0,$$

here $\Omega^0(R^3)$ is the space of smooth functions on R^3 ,$\Omega^1(R^3)$ is the space of 1 forms, i.e. $\Omega^i(R^3)$ is the space of i forms. Then we can choose E as a differential form of u of degree 1, H as a differential form of f of degree 2. More explicitly, we identify the vector $E=(E_1,E_2,E_3)$ with the 1-form $u=E_1dx^1+E_2dx^2+E_3dx^3$ and the vector $H=(H_1,H_2,H_3)$ with the 2-form f= H_1 ★ + ★ + dx^1 H_2 dx^2 H_3 dx^3 ★where ★ is the Hodge operator. Then $rotE$ exactly corresponds to du and $rotH$ with d^*f , where d^* =★ d ★ is the formally conjugate operator for d . Using the operators $d:\Omega^i \to \Omega^{i+1}$ (the outer differentiation operator increasing the degree of the form by one) and $d^*:\Omega^i \to \Omega^{i-1}$ (the operator d^* decreasing the degree of the form by one) we construct the well-known operator or, otherwise, the de Rham Laplacian $\Delta = d^*d + dd^*$, which is a formally self-adjoint non-negative elliptic differential operator of order two $\Delta:\Omega^i(X) \to \Omega^i(X)$.

The operator d^* is conjugate to the operator external with respect to the scalar product

$\langle du, f\rangle = \langle u, d^* f\rangle$, where $u \in \Omega^i(X)$ и $f \in \Omega^{i+1}(X)$.

The operator Δ commutes with the two original operators $d\Delta = \Delta d$ and $d^*\Delta = \Delta d^*$, and the Hodge operator ★$\Delta = \Delta$ ★ , which is a well-known isomorphism of vector spaces, ★: $\Omega^i(X) \to \Omega^{i-1}(X)$, i.e.

$$\bigstar\ dx^i = (-1)^{i-1} dx^1 \wedge dx^2 \wedge \ldots \wedge \hat{d}^i x \wedge dx^{i+1} \wedge \ldots \wedge dx^n$$

$$\bigstar dx^1 = dx^2 \wedge dx^3 \;,\; \bigstar dx^2 = dx^3 \wedge dx^1 \;,\; \bigstar dx^3 = dx^1 \wedge dx^2 \qquad \text{to } R^3 \;.$$

For 1-form:

$$du = d(E_1dx^1 + E_2dx^2 + E_3dx^3) = \left(\frac{\partial E_1}{\partial x^1}dx^1 + \frac{\partial E_1}{\partial x^2}dx^2 + \frac{\partial E_1}{\partial x^3}dx^3\right)dx^1 +$$

$$+\left(\frac{\partial E_2}{\partial x^1}dx^1 + \frac{\partial E_2}{\partial x^2}dx^2 + \frac{\partial E_2}{\partial x^3}dx^3\right)dx^2 + \left(\frac{\partial E_3}{\partial x^1}dx^1 + \frac{\partial E_3}{\partial x^2}dx^2 + \frac{\partial E_3}{\partial x^3}dx^3\right)dx^3$$

$$= \left(\frac{\partial E_3}{\partial x^2} - \frac{\partial E_2}{\partial x^3}\right)dx^2 \wedge dx^3 - \left(\frac{\partial E_1}{\partial x^3} - \frac{\partial E_3}{\partial x^1}\right)dx^1 \wedge dx^3 + \left(\frac{\partial E_2}{\partial x^1} - \frac{\partial E_1}{\partial x^2}\right)dx^1 \wedge dx^2$$

$$= \left(\frac{\partial E_3}{\partial x^2} - \frac{\partial E_2}{\partial x^3}\right)dx^2 \wedge dx^3 + \left(\frac{\partial E_1}{\partial x^3} - \frac{\partial E_3}{\partial x^1}\right)dx^3 \wedge dx^1 + \left(\frac{\partial E_2}{\partial x^1} - \frac{\partial E_1}{\partial x^2}\right)dx^1 \wedge dx^2$$

Для 2 -forms:

$$d^* f \quad = \bigstar d \;\bigstar f \;= \bigstar d(H_1dx^1 + H_2dx^2 + H_3dx^3) =$$

$$\bigstar\left[\left(\frac{\partial H_3}{\partial x^2} - \frac{\partial H_2}{\partial x^3}\right)dx^2 \wedge dx^3 + \left(\frac{\partial H_1}{\partial x^3} - \frac{\partial H_3}{\partial x^1}\right)dx^3 \wedge dx^1 + \left(\frac{\partial H_2}{\partial x^1} - \frac{\partial H_1}{\partial x^2}\right)dx^1 \wedge dx^2\right] =$$

$$= \left(\frac{\partial E_3}{\partial x^2} - \frac{\partial E_2}{\partial x^3}\right)dx^1 + \left(\frac{\partial E_1}{\partial x^3} - \frac{\partial E_3}{\partial x^1}\right)dx^2 + \left(\frac{\partial E_2}{\partial x^1} - \frac{\partial E_1}{\partial x^2}\right)dx^3$$

Thus, the system of Maxwell's equations (1.1.1) can be written in the form

$$\begin{aligned} &iku + d^* f = 0, \\ &-ikf + du = 0, \end{aligned} \qquad (2.1.1)$$

which makes sense not only for the differential forms u and f of degree 1 and 2 in R^3 , but also for the differential forms u and f of degree i and $\mathrm{i}+1$ respectively in R^n , where $.-1 \le \mathrm{i} \le n$

Definition 2.1. *Let* $-1 \le \mathrm{i} \le 3$ *. We define the Maxwell operator for the de Rham complex in* R^3 *at step* i *as*

$$M^{\mathrm{i}} = \begin{pmatrix} ik & d^{\mathrm{i}*} \\ d^{\mathrm{i}} & -ik \end{pmatrix} \qquad .$$

On the other hand, applying d^* to both sides from $iku + d^* f = 0$, we conclude that $d^{i-1^*} u = 0$, if $k \neq 0$. Similarly, it follows from $-ikf + du = 0$, that $d^{i+1} f = 0$ if $. k \neq 0$

A characteristic and important example of a system elliptic by Duglis-Nirenberg and non-elliptic by Petrovsky, is the system of Maxwell's equations (2.1.1) in the domain $. X \subset R^3$

Example 2.1. *In the case of the de Rham complex in* R^3 *at step* i =0 *the system of Maxwell's equations (2.1.1) has the following form*

$$iku - divf = 0,$$
$$-ikf + gradu = 0,$$

or

$$\begin{aligned} -iku + divf &= 0, \\ -ikf + gradu &= 0. \end{aligned} \qquad (2.1.2)$$

The system (2.1.2) with constant coefficients and its operator M^0 is:

$$M^0 = \begin{pmatrix} -ik & \frac{\partial}{\partial x_1} & \frac{\partial}{\partial x_2} & \frac{\partial}{\partial x_3} \\ \frac{\partial}{\partial x_1} & -ik & 0 & 0 \\ \frac{\partial}{\partial x_2} & 0 & -ik & 0 \\ \frac{\partial}{\partial x_3} & 0 & 0 & -ik \end{pmatrix}$$

Let us try to find vectors s and t with integer components that simultaneously satisfy conditions (1.1.9) , (1.1.11).

For the system (2.1.2) we choose the sets (1.1.9), e.g., the form

$s = (2,1,1,1)$ and $. t = (0,-1,-1,-1)$

The principal part of the operator M^0 corresponding to (1.1.10) has the form

$$\widetilde{M}^0_{s,t}(D)=\begin{pmatrix} 0 & \frac{\partial}{\partial x_1} & \frac{\partial}{\partial x_2} & \frac{\partial}{\partial x_3} \\ \frac{\partial}{\partial x_1} & -ik & 0 & 0 \\ \frac{\partial}{\partial x_2} & 0 & -ik & 0 \\ \frac{\partial}{\partial x_3} & 0 & 0 & -ik \end{pmatrix},$$

and its determinant

$$\det \widetilde{M}^0_{s,t}(D)=k^2(\xi_1^2+\xi_2^2+\xi_3^2)\neq 0 \qquad \forall \xi\in R^3, k\neq 0$$

Using Definition 2.1, we conclude that the system (2.1.1) is Douglas-Nirenberg elliptic.

Example 2.2. *In the case of the de Rham complex in* R^2 *at step* i =*2 the Maxwell operator* M^i *has the form*

$$M^2=\begin{pmatrix} -ik & 0 & \frac{\partial}{\partial x_1} \\ 0 & -ik & \frac{\partial}{\partial x_2} \\ \frac{\partial}{\partial x_1} & \frac{\partial}{\partial x_2} & -ik \end{pmatrix}.$$

For sets satisfying condition (1.1.9) $s=(0,0,1)$, , $t=(0,0,1)$ relevant main part

$$\widetilde{M}^2_{s,t}(D)=\begin{pmatrix} -ik & 0 & \frac{\partial}{\partial x_1} \\ 0 & -ik & \frac{\partial}{\partial x_2} \\ \frac{\partial}{\partial x_1} & \frac{\partial}{\partial x_2} & 0 \end{pmatrix}$$

it corresponds to the determinant:

$$\det \widetilde{M}^2_{s,t}(D) = \begin{vmatrix} -ik & 0 & \xi_1 \\ 0 & -ik & \xi_2 \\ \xi_1 & \xi_2 & 0 \end{vmatrix} = ik(\xi_1^2 + \xi_2^2) \neq 0, \ \forall \xi \in \mathrm{R}^2 \setminus \{0\}, k \neq 0.$$

Therefore, the system (2.1.1) in the case of the de Rham complex in R^2 at step =2 i

will be elliptical according to Douglas-Nirenberg.

Example 2.3. *In the case of the de Rham complex in* R^n *at step* i *=2 the Maxwell operator* M^i *has the form*

$$M^2 = \begin{pmatrix} -ik & 0 & 0 & 0 & \dots & \frac{\partial}{\partial x_1} \\ 0 & -ik & 0 & 0 & 0 & \frac{\partial}{\partial x_2} \\ 0 & 0 & -ik & 0 & 0 & \frac{\partial}{\partial x_3} \\ 0 & 0 & 0 & -ik & 0 & \vdots \\ \vdots & 0 & 0 & 0 & \ddots & \frac{\partial}{\partial x_n} \\ \frac{\partial}{\partial x_1} & \frac{\partial}{\partial x_2} & \frac{\partial}{\partial x_3} & \dots & \frac{\partial}{\partial x_n} & -ik \end{pmatrix}$$

corresponding to the sets $s = (s_1 = 0, s_2 = 0, \dots, s_{N+1} = 1)$, $t = (t_1 = 0, t_2 = 0, \dots, t_{N+1} = 1)$, the main part of the matrix differential operator M^i

$$\widetilde{M}^2_{s,t} = \begin{pmatrix} -ik & 0 & 0 & 0 & \dots & \frac{\partial}{\partial x_1} \\ 0 & -ik & 0 & 0 & 0 & \frac{\partial}{\partial x_2} \\ 0 & 0 & -ik & 0 & 0 & \frac{\partial}{\partial x_3} \\ 0 & 0 & 0 & -ik & 0 & \vdots \\ \vdots & 0 & 0 & 0 & \ddots & \frac{\partial}{\partial x_n} \\ \frac{\partial}{\partial x_1} & \frac{\partial}{\partial x_2} & \frac{\partial}{\partial x_3} & \dots & \frac{\partial}{\partial x_n} & 0 \end{pmatrix}$$

and its determinant

$$\det \widetilde{M}^0_{s,t}(D) \neq 0 \qquad \forall \xi \in R^n, k \neq 0$$

Therefore, the system (2.1.1) in the case of the de Rham complex in R^n at step i =2 will be Douglas-Nirenberg elliptic.

As usual, we write $M^i(u,f)$ simply $M(u,f)$ for $u \in C^\infty(X,F^i)$, when there can be no confusion.

To study the Cauchy problem, we need some integral representation. Usually it is constructed by means of a suitable fundamental solution and Green's formula. To circumvent this difficulty, one can, for example, choose the operator

$$C = \begin{pmatrix} ik + (1/ik)dd^* & d^* \\ d & -ik - (1/ik)d^*d \end{pmatrix} \qquad (2.1.3)$$

in a way that would be fair

Lemma 2.1. *As defined above,* C *satisfies*

$$CM = MC = \begin{pmatrix} \Delta - k^2 & 0 \\ 0 & \Delta - k^2 \end{pmatrix}.$$

Proof.

$$CM = MC =$$

$$= \begin{pmatrix} -k^2 + dd^* + d^*d & ikd^* + (1/ik)dd^*d^* - ikd^* \\ ikd - ikd - (1/ik)d^*dd & dd^* - k^2 + d^*d \end{pmatrix} =$$

$$= \begin{pmatrix} \Delta - k^2 & 0 \\ 0 & \Delta - k^2 \end{pmatrix}.$$

Let us denote by G the fundamental solution for $\Delta - k^2$.

Let X be a compact manifold with edge smoothly embedded in a larger one

C^∞ manifold X' , and the de Rham sequence is defined over the whole $X' \subset R^3$. If $X' = R^3$, , then the image for X can be any bounded region in the R^3

Lemma 2.2. *Pseudodifferential operator*

$$\Phi = \begin{pmatrix} G(ik + (1/ik)dd^*) & Gd^* \\ Gd & G(-ik - (1/ik)d^*d) \end{pmatrix}$$

there is a left fundamental solution of the Maxwell operator M *to* X' .

Proof . It immediately follows from lemma 2.1 that

$$\Phi = \begin{pmatrix} G & 0 \\ 0 & G \end{pmatrix} \circ C$$

is the left fundamental solution of M . It remains to replace the explicit expression (1.1.3) for $.C$

Theorem 2.2. *For every* $u \in H^1(X, F^i)$ *and* $f \in H^1(X, F^{i+1})$ *, satisfying* $M^i(u, f) = 0$ *inside* X *, it follows that*

$$\int_{\partial X} \Phi(x,y)\begin{pmatrix} in(f) \\ -\nu \wedge t(u) \end{pmatrix} ds = \begin{cases} \begin{pmatrix} u(x) \\ f(x) \end{pmatrix}, & \text{если} \quad x \in X \setminus \partial X, \\ 0, & \text{если} \quad x \in X' \setminus X, \end{cases}$$

here, $\nu(y)$ is the outer unit vector of the normal to the boundary at the point $y \in \partial X$, ds surface measure at . X

Proof. This formula is a special case of the general Green's formula [37, 2.5.4], where

$$\sigma(M) = \begin{pmatrix} 0 & \sigma(d^*) \\ \sigma(d) & 0 \end{pmatrix}.$$

For the classical Maxwell equations, the formula of Theorem 2.1 is known as the Stratton-Chu formula, see [36]. [36]. In this case it exhibits a more refined structure, since the fundamental solution Φ can be written explicitly.

Corollary 2.1. *Let* (E, H) *be an electromagnetic wave in* X, *the electric component* E *and the magnetic component* H *, which is continuous up to the boundary. Then*

$$\begin{pmatrix} (1/ik)d^*d & -d^* \\ -d & -(1/ik)dd^* \end{pmatrix} \int_{\partial X} e(x-y)\begin{pmatrix} in(H) \\ -\nu \wedge t(E) \end{pmatrix} ds = \begin{pmatrix} E(x) \\ H(x) \end{pmatrix}$$

for all $x \in X \setminus \partial X$ *, and the left side disappears further from .* X

Proof . From (2.1.3) it is easy to deduce that.

$$C=\begin{pmatrix}\frac{1}{ik}((ik)^2+dd^*) & d^* \\ d & -\frac{1}{ik}((ik)^2+d^*d)\end{pmatrix}=$$

$$\begin{pmatrix}\frac{1}{ik}(\Delta-k^2-d^*d) & d^* \\ d & -\frac{1}{ik}(\Delta-k^2-dd^*)\end{pmatrix},$$

so

$$\Phi(x,y)=-\begin{pmatrix}-\frac{1}{ik}d^*de(x-y) & d^*e(x-y) \\ de(x-y) & \frac{1}{ik}dd^*e(x-y)\end{pmatrix}$$

in $R^3\setminus\{y\}$ for $(\Delta+k^2)=0$ away from the origin.

To complete the proof, it suffices to use Theorem 2.1 and notice that $\sigma(\Delta)(v)=|v|^2 I_{k_3}=I_{k_3}$ in our special case.

It can easily be verified that

$$M\begin{pmatrix}\frac{1}{ik}d^*d & -d^* \\ -d & -\frac{1}{ik}dd^*\end{pmatrix}=\begin{pmatrix}0 & -\frac{1}{ik}d^*(\Delta-k^2) \\ \frac{1}{ik}d(\Delta-k^2) & 0\end{pmatrix},$$

hence, the left part of the formula in Corollary 2.1 satisfies Maxwell's equations in $R^3\setminus\partial X$ for all integrable functions $t(u)$ and $n(f)$ on the boundary. Thus, we arrive at what is usually called the Cauchy type integral associated with Maxwell's equations, [37, 3.2.3].

§2.2 The Cauchy problem. System with double orthogonality

Combining the systems of Maxwell's equations with the data from $t(u)$ and $n(f)$ we determine u and f at the boundary, for the

$$n(u) = -(1/ik)n(d^* f),$$
$$t(f) = (1/ik)t(du)$$

$n(d^* f)$ and $t(du)$ is uniquely determined by $n(f)$ and $t(u)$ (the so-called tangent operators on ∂X , see[37, 3.1.5]). Thus, the Cauchy problem for the system of Maxwell's equations in X consists of finding the solution of (u, f) in $M(u, f) = 0$ with data $t(u)$ and $n(f)$ on a part of the surface ∂X [39],[13].

Let S be an open part on the boundary of X . The Cauchy problem for the system of Maxwell's equations in X with data at S is as follows. Find a solution of (u, f) satisfying $M(u, f) = 0$ inside X and such that $t(u) = u_0$ and $n(f) = f_0$ on : S

$$\begin{cases} M(u, f) = 0 & в \quad X \setminus \partial X, \\ t(u) = u_0 & на \quad S, \\ n(f) = f_0 & на \quad S. \end{cases}$$

(2.2.1)

To study problem (2.2.1), we introduce the integral completely defined by the Cauchy data u_0 and f_0 to S , namely

$$G(u_0, f_0)(x) = \int_S \Phi(x-y) \begin{pmatrix} if_0 \\ -\nu \wedge u_0 \end{pmatrix} ds$$

for $x \in R^3 \setminus \overline{S}$. Since the fundamental solution of $\Phi(x-y)$ is real analytic off-diagonal to $R^3 \times R^3$, then $G(u_0, f_0)$ is real analytic in addition to . $\overline{S}$

Moreover, the potential $G(u_0, f_0)$ satisfies Maxwell's equations $M(u, f) = 0$ in $R^3 \setminus \overline{S}$. In particular, the components of the vector-function $G(u_0, f_0)$ turn out to be solutions of the (scalar) Helmholtz equation $(\Delta + k^2)G = 0$ in the complement of $\overline{S}$.

Theorem 2.2. *For there to exist a solution* (u,f) *of the Cauchy problem (2.2.1) continuous up to* S *, it is necessary and sufficient that the integral* $G(u_0,f_0)$ *can be continued from* $R^3\setminus X$ *through* S *to* X *as a real-analytic function.*

Proof. *Necessity.* Suppose that there exists a solution (u,f) of the Cauchy problem (2.2.1) which is continuous to S . Let us predefine the electric and magnetic fields U и F

$$\begin{pmatrix} U(x) \\ F(x) \end{pmatrix} = \begin{cases} G(x_0,y_0) - \begin{pmatrix} u \\ f \end{pmatrix} & в \ X\setminus\partial X, \\ G(u_0,f_0) & в \ R^3\setminus X. \end{cases}$$

Let us write $U^{\pm}$ for the contraction of U into open sets $X\setminus\partial X$ and $R^3\setminus\overline{X}$, respectively, and similarly for F . Reducing X , if necessary, we can assume that u and f are continuous up to the boundary of X . Using Theorem 2.1 [43], we obtain

$$\begin{pmatrix} U^+(x) \\ F^+(x) \end{pmatrix} = \int_S \Phi(x-y)\begin{pmatrix} if_0 \\ -\nu\wedge u_0 \end{pmatrix} ds - \int_{\partial X} \Phi(x,y)\begin{pmatrix} in(f) \\ -\nu\wedge t(u) \end{pmatrix} ds$$
$$= - \int_{\partial X\setminus S} \Phi(x-y)\begin{pmatrix} in(f) \\ -\nu\wedge t(u) \end{pmatrix} ds$$

for all x inside the neighborhood of X . It follows that U^+ and F^+ continue through S as real-analytic functions in the whole $(R^3\setminus\partial X)\cup S$. Again applying Theorem 2.1 of [46] we have

$$- \int_{\partial X\setminus S} \Phi(x,y)\begin{pmatrix} in(f) \\ -\nu\wedge t(u) \end{pmatrix} ds = \int_S \Phi(x,y)\begin{pmatrix} in(f) \\ -\nu\wedge t(u) \end{pmatrix} ds = \begin{pmatrix} U^-(x) \\ F^-(x) \end{pmatrix}$$

where $x\in R^3\setminus X$. Hence, $G(u_0,f_0)$ continues in $R^3\setminus X$ through S inside X as a real analytic function, which was required to prove.

Sufficiency. Conversely, let U and F be real analytic on $(R^3 \setminus \partial X) \cup S$, such that

$$\begin{pmatrix} U(x) \\ F(x) \end{pmatrix} = G(u_0, f_0)$$

away from X. Then $M(U,F) = 0$ is in $R^3 \setminus X$. Since $M(U,F)$ is real analytic in $(R^3 \setminus \partial X) \cup S$, in fact it is also zero in . X

Let's

$$\begin{pmatrix} u(x) \\ f(x) \end{pmatrix} = G(u_0, f_0)(x) - \begin{pmatrix} U(x) \\ F(x) \end{pmatrix}$$

where x is inside the neighborhood of X. From what has already been proved that u and f are continuous up to S and satisfy Maxwell's equation. Let us show that, (u, f) is the desired solution of (2.2.1). To verify this, it remains to check that $t(u) = u_0$ and $n(f) = f_0$ are at S. For this purpose, let us denote by $G_e(u_0, f_0)$ and $G_m(u_0, f_0)$ the components of $G(u_0, f_0)$, which are the electric and magnetic fields, respectively. Since U and F are continuous, in $(R^3 \setminus \partial X) \cup S$, by the Sohotsky-Plemel formula we have [9, p. 105]

$$\begin{aligned} t(u) &= t(G_e(u_0, f_0)^+ - U^+) \\ &= t(G_e(u_0, f_0)^+ - U^-) \\ &= t(G_e(u_0, f_0)^+ - G_e(u_0, f_0)^-) \\ &= u_0 \end{aligned}$$

at S. Similarly,

$$\begin{aligned} n(f) &= n(G_m(u_0, f_0)^+ - F^+) \\ &= n(G_m(u_0, f_0)^+ - F^-) \\ &= n(G_m(u_0, f_0)^+ - G_m(u_0, f_0)^-) \\ &= f_0 \end{aligned}$$

at S, which is our assertion.

Theorem 2 can be generalized by saying that any explicit formula for the analytic continuation from $R^3 \setminus X$ through S to the interior of X leads to a formula for solutions to the Cauchy problem with data at S . This idea goes back at least to [3],[19].

Systems with double orthogonality

The problem of continuation in Hilbert space has a satisfying solution in terms of a basis with double orthogonality [12],[30] ,[14],[38](see also [7] for more general systems). This idea belongs to S. Bergman (1927), who used it to derive criteria for analytic continuation.

We will apply this method to the solution of the Cauchy problem for a homogeneous system of Maxwell's equations in the special case where X is part of a ball $B = B(0,R)$ with center (at the origin) at zero and radius, $R > 0$ before developing the theory in the general case. Let S be a smooth closed hypersurface, in B, which does not pass through the point $x = 0$ and splits B into two regions. Let us denote by X the point that does not contain the origin. Its boundary consists of the point S and the part of the sphere ∂B in R^3 cm. Figure 1. The advantages of using the region X specified is that the problem reduces to the analytic continuation of a small sphere around 0 to B . As mentioned, the integral of $G(u_0, f_0)$ satisfies both equations $MG(u_0, f_0) = 0$ and $(\Delta + k^2)G(u_0, f_0) = 0$ outside the closure of X . The last scalar Helmholtz equation follows from the first one. Here we use a basis with double orthogonality in Hilbert space, the solution of the Helmholtz equation to derive the condition for analytic continuation from $B(0,\varepsilon)$ to $B(0,R)$ for

$G(u_0, f_0)$. The Helmholtz operator $\Delta + k^2$ in R^3 spherical coordinate space is as follows:

$$\Delta + k^2 = \frac{1}{r^2}\left(\left(r\frac{\partial}{\partial r}\right)^2 + r\frac{\partial}{\partial r} + k^2 r^2 - \Delta_S\right),$$

(2.2.2)

where $\Delta_S = -\frac{1}{\sin\theta}\frac{\partial}{\partial\theta}\left(\sin\theta\frac{\partial}{\partial\theta}\right) - \frac{1}{\sin^2\theta}\frac{\partial^2}{\partial\varphi^2}$ is the Laplace-Beltrami operator on the unit sphere. Recall that k^2 is an arbitrary real number.

To solve the homogeneous equation $(\Delta + k^2)u = 0$ we will use the Fourier method, i.e., the method of separation of variables. Let's put

$$u = g(r,k)h(\varphi)$$

we obtain two separate equations for g and h

$$\left(\left(r\frac{\partial}{\partial r}\right)^2 + r\frac{\partial}{\partial r} + k^2 r^2\right)g = cg,$$

$$\Delta_S h = ch,$$

where c is an arbitrary constant.

The second equation has a non zero solution when c is an eigenvalue of the operator Δ_S . These eigenvalues are well known $c = \nu(\nu+1)$, where $\nu = 0,1,...,$ [40].

The eigenfunctions corresponding to these eigenvalues Δ_S are spherical harmonics $h_\nu(\varphi)$ of degree ν , i.e.

$$\Delta_S h_\nu = \nu(\nu+1)h_\nu.$$

(2.2.3)

Now consider a homogeneous differential equation with respect to the variable $r > 0$

$$\left(\left(r\frac{\partial}{\partial r}\right)^2 + r\frac{\partial}{\partial r} + (k^2r^2 - \nu(\nu+1))\right)g(r,k) = 0. \qquad (2.2.4)$$

This is the Bessel equation and its solution space is two-dimensional. For example, if $k = 0$, then $g(r,0) = ar^{\nu} + br^{-\nu-1}$ with arbitrary constants a and b is a general solution of (2.2.4). In this case the function $r^{\nu}h_{\nu}(\varphi)$ is a homogeneous harmonic polynomial. In the general case, the space of solutions of (2.2.4) contains a one-dimensional subspace of functions bounded at the point $r = 0$, [40].

For $\nu = 0,1,...,$ a fixed non zero solution $g_{\nu}(r,k)$ of equation (1.22), which is bounded at $r = 0$. Then

$$(\Delta + k^2)(g_{\nu}(r,k)h_{\nu}(\varphi)) = 0 \qquad (2.2.5)$$

in the whole R^3 . Indeed, we conclude from (2.2.2.2), (2.2.3) and (2.2.4) that this equality holds in $R^3 \setminus \{0\}$. Now noticing that $g_{\nu}(r,k)h_{\nu}(\varphi)$ is bounded, we first make sure that (2.2.5) is true.

It is known that the number of linearly independent spherical harmonics in degree ν is equal to $J(\nu) = 2\nu + 1$. Let us take an orthonormal basis

$$\{h_{\nu}^{(j)}\}_{\substack{\nu=0,1,...\\ j=1,...,J(\nu)}} \qquad \text{в } L^2(S).$$

Lemma 2.3. *For each* $R > 0$, *system*

$$\{b_{\nu}^{j}(r,\varphi,k)\} := g_{i}(r,k)h_{\nu}^{j}(\varphi)_{\substack{\nu=0,1,...\\ j=1,...,J(\nu)}} \qquad (2.2.6)$$

is an orthogonal basis in the subspace $L^2(B(0,R))$, *consisting of solutions of Helmholtz equations* $(\Delta + k^2)u = 0$.

Proof. Indeed, $\{h_\nu^{(j)}\}$ an orthonormal basis in the space $L^2(S)$ on the unit sphere (2.2.6) is orthogonal in $L^2(B(0,R))$, because

$$\left(b_\nu^{(j)}, b_l^m\right)_{L^2(B(0,R))} = \left(h_\nu^j, h_l^m\right)_{L^2(S)} \int_0^R g_\nu(r,k)\overline{g_l(r,k)} \cdot r^2 dr = 0$$

where $\left(h_\nu^j, h_l^m\right)_{L^2(S^1)} = \int_0^1 h_\nu^j(\varphi)\overline{h_l^m(\varphi)}d\varphi = \delta_{\nu l}^{jm} = 0$

for $\nu \neq l$ or $. j \neq m$ Finally, since the system of $\{h_\nu^{(j)}\}$ harmonics is dense in $C^\infty(S)$ we see the system (2.2.6) is dense in $L^2(B(0,R)) \cap S_{\Delta+k^2}(B(0,R))$, this is the proof.

We can take the standard kernel as the fundamental solution of the operator $\Delta + k^2$ in R^3

$$\frac{-1}{4\pi}\frac{exp(ik|x-y|)}{|x-y|}.$$

For a given fixed $y \in R^3 \setminus \{0\}$, this kernel is a solution of the Helmholtz equation $(\Delta + k^2)u = 0$ in the ball $B(0,|y|)$. Obviously, it is quadratically integrable in $B(0,|y|)$, and we denote by $c_\nu^{(j)}(y,k)$ its Fourier coefficients with respect to the orthogonal system (2.2.6), that is,

$$c_\nu^{(j)}(y,k) = \left(\frac{-1}{4\pi}\frac{exp(ik|x-y|)}{|x-y|}, b_\nu^{(j)}(x,k)\right)_{L^2(B(0,|y|))} / \int_0^{|y|} |g_\nu(r,k)|^2 r^2 dr$$

Lemma 2.4. *In the cone* $\{(x,y) \in R^3 \times R^3 : |x|/|y| < 1\}$ *there is a Fourier series expansion*

$$\frac{-1}{4\pi}\frac{exp(ik|x-y|)}{|x-y|} = \sum_{\nu=0}^{\infty}\sum_{j=1}^{J(\nu)} c_\nu^j(y,k) b_\nu^j(x,k), \tag{2.2.7}$$

where the series converges uniformly with all derivatives on compact subsets of the cone.

Proof. The expansion into Fourier series is a direct consequence of Lemma 2.3. The uniform convergence on compact sets is a more delicate question. It can be obtained by the well-known method for $b_\nu^{(j)}$ being a solution of the Helmholtz equation.

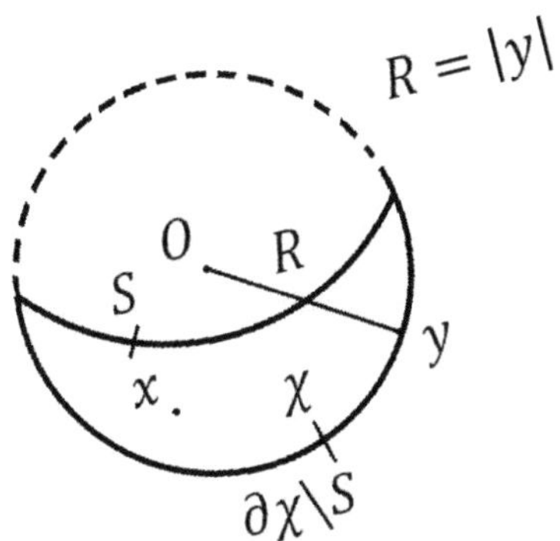

Figure 1.

§2.3 Carleman's formula for a part of a ball

Substituting (2.2.7) into the formula for the fundamental solution of $\Phi(x-y)$ for M , we obtain

$$\Phi(x-y)=\sum_{\nu=0}^{\infty}\Phi_\nu(x,y) \tag{2.3.1}$$

where the series converges uniformly together with all its derivatives on compact subsets of the cone $\{(x,y)\in R^3\times R^3 : |x|/|y|<1\}$ and

$$\Phi_\nu(x,y) = \begin{pmatrix} (1/ik)d^*d & -d \\ -d & -(1/ik)dd^* \end{pmatrix} \sum_{j=1}^{J(\nu)} c_\nu^{(j)}(y,k) b_\nu^{(j)}(x,k)$$

for $\nu = 0,1,...$. We get the same formula if we substitute together $-\partial_y$ for the derivatives of . ∂_x

Lemma 2.5. *Each member of* $\Phi_\nu(x,y)$ *is a real-analytic matrix-valued function on* $R^3 \times (R^3 \setminus \{0\})$ *, satisfying*

$$M(\partial_x)\Phi_\nu(x,y) = 0,$$
$$M'(\partial_y)(\Phi_\nu(x,y))^T = 0.$$

Proof. These properties are obvious by the very construction of degenerate kernels $\Phi_\nu(x,y)$. The singularity at $y = 0$ is due to the integral

$$\int_0^{|y|} |g_\nu(r,k)|^2 r^2 dr \qquad .$$

A series expansion similar to (2.3.1) with terms Φ_ν satisfying the transposed equation $M'(\partial_y)(\Phi_\nu(x,y))^T = 0$ is already enough to obtain an explicit formula for solutions of the Cauchy problem (2.2.1).

Let's

$$R_N(x,y) = \Phi(x-y) - \sum_{\nu=0}^{N} \Phi_\nu(x,y)$$

for $(x,y) \in R^3 \times (R^3 \setminus \{0\})$. Thus, we obtain the so-called Carleman function solution of the Cauchy problem, see. [39, 10.4],[6], [12].

Theorem 2.3. *Suppose that* (u,f) *is a solution of Maxwell's equations inside* X *, continuous up to the closure of* S *. Then*

$$\begin{pmatrix} u(x) \\ f(x) \end{pmatrix} = \lim_{N\to\infty} \int_S R_N(x,y) \begin{pmatrix} in(f) \\ -\nu \wedge t(u) \end{pmatrix} ds$$

for everyone . $x \in X \setminus \partial X$

Proof. We can assume, by approximating X by smaller regions whose boundaries intersect the boundary of X only in S , that (u,f) is smooth in X . Combining Corollary 2.1 (Chapter 2) with Stokes' formula and Lemma 2.5, we easily conclude that

$$\begin{pmatrix} u(x) \\ f(x) \end{pmatrix} = \int_{\partial X} R_N(x,y) \begin{pmatrix} in(f) \\ -\nu \wedge t(u) \end{pmatrix} ds$$

for each fixed $x \in X \setminus \partial X$ and all $N = 0,1,...$. Let $N \to \infty$. Since the series (2.3.1) converges on $\partial B(0,R)$ uniformly over x on compact subsets of $B(0,R)$, it follows that the part of the boundary integral over $\partial X \setminus S$ tends to zero. This establishes the desired formula.

Let (u,f) be the solution of the Cauchy problem (2.2.1). A simple calculation shows that

$$\int_S R_N(x,y) \begin{pmatrix} in(f) \\ -\nu \wedge t(u) \end{pmatrix} ds = G(u_0,f_0)(x) - \begin{pmatrix} U_N(x) \\ F_N(x) \end{pmatrix}$$

(2.3.2)

for $x \notin \overline{S}$, where

$$\begin{pmatrix} U_N(x) \\ F_N(x) \end{pmatrix} = \sum_{\nu=0}^{N} \int_S \Phi_\nu(x,y) \begin{pmatrix} in(f) \\ -\nu \wedge t(u) \end{pmatrix} ds .$$

Let $\varepsilon > 0$ be the distance between S and the origin. If $x \in B(0,\varepsilon)$, then the left part of (2.3.2) tends to zero, since the series (2.3.1) converges uniformly to S . It follows that the series (i.e., the sequence of partial sums) $\{(U_N, F_N)\}$ converges to $G(u_0, f_0)$ uniformly together with its derivatives on compact subsets of the ball . $B(0,\varepsilon)$

The last observation and Theorem 2.1 combine to give some solvability conditions for the Cauchy problem.

Corollary 2.2. *If the series* $\{(U_N, F_N)\}$ *converges uniformly on compact subsets of the ball* $B(0,R)$ *, then the Cauchy problem (2.2.1) is solvable.*

Proof. Since the terms of the series $\{(U_N, F_N)\}$ are component solutions of the Helmholtz equation, it follows from the Stiltjes-Vitali theorem that its sum $(U,F) = \lim(U_N, F_N)$ satisfies in components the same equation in $B(0,R)$. Hence, (U,F) is the real analytic cross section of $F^i \oplus F^{i+1}$ on $B(0,R)$. Since (U,F) is effectively the same as $G(u_0, f_0)$ in the smaller ball $B(0,\varepsilon)$, the solvability of the Cauchy problem follows from Theorem 2.1. In many interesting cases, uniform convergence of the series $\{(U_N, F_N)\}$ is not only sufficient but also necessary for the solvability of (5.1), [35, 2.9], [14].

As mentioned in paragraph 1.1, classically the system of Maxwell's equation (1.1.1) has the form:

$$\begin{aligned} ikE + d^* H &= 0, \\ -ikH + dE &= 0, \end{aligned}$$

where E and H are functions in a closed region $X \subset R^3$ with values in R^3 . If E is introduced in the specified limits by 1-form and H by 2-form and both the outer derivative d and the formally conjugate to it d^* can be identified with the operator rot in vector fields in . R^3

Applying Corollary 3.3 of [45],[46] to the system of classical Maxwell's equations we obtain the Stratton-Chu formula.

Theorem 2.4. *Let* (E,H) *be an electromagnetic wave in* X *, whose electric component* E *and magnetic component* H *are continuous up to the boundary. Then*

$$\begin{pmatrix}(1/ik)d^*d & -d^* \\ -d & -(1/ik)dd^*\end{pmatrix}\int_{\partial X}\frac{-1}{4\pi}\frac{exp(ik\,|x-y|}{|x-y|}\begin{pmatrix}in(H) \\ -\nu\wedge t(E)\end{pmatrix}ds=\begin{pmatrix}E(x) \\ H(x)\end{pmatrix}$$

for all $x \in X\setminus\partial X$ *, and the left side disappears further from* . X

Let X the region in R^3 and S the open part at ∂X . We consider the problem of finding the electric field E and magnetic field H in X with given tangential component E_0 in E and normal component H_0 in H at . S

Thus, we obtain the so-called Carleman function for solutions of the Cauchy problem. Namely

$$K_N(x,y)=\begin{pmatrix}(1/ik)d^*d & -d^* \\ -d & -(1/ik)dd^*\end{pmatrix}\left(\frac{-1}{4\pi}\frac{exp(ik|x-y|)}{|x-y|}-\sum_{\nu=0}^{N}\sum_{j=1}^{J(\nu)}c_\nu^j(y,k)b_\nu^j(x,k)\right)$$

,

differential operator on the right-hand side acting on the variable x .

Theorem 2.5. *Let* (E,H) *electromagnetic waves in* X *continuous up to* $\overline{S}$ *. Then*

$$\begin{pmatrix}E(x) \\ H(x)\end{pmatrix}=\lim_{N\to\infty}\int_S K_N(x,y)\begin{pmatrix}in(H) \\ -\nu\wedge t(E)\end{pmatrix}ds$$

for everyone . $x \in X\setminus\partial X$

In [39], several approaches to the Cauchy problem for solutions of linear elliptic equations with data on a part of the boundary are developed. They give not only explicit Carleman-type formulas for the solutions, but also conditions for the Cauchy data that are necessary and sufficient to solve the Cauchy problem. The results rather strikingly show that the Cauchy problem for elliptic equations is overdetermined.

It is solvable for a fine set of Cauchy data, and so any error in the Cauchy data leads to unsolvability. For this reason, the variational approach to the Cauchy problem for elliptic equations is of practical interest. It works also in the case of nonlinear elliptic equations, see. [29]. More generally, the problem is to introduce reasonable classes of approximate solutions. In this sense, any Carleman formula, such as the formula of Theorem 2.5, gives approximate solutions to the Cauchy problem with data on the boundary segment for Maxwell's equation.

§2.4 Carleman's formula in the cap-type domain
Fundamental lemma

However, no perfect proof was presented until the paper [24] appeared, where the formula obtained in [17] was generalized to the case of the Laplace equation. Such a proof was first obtained in [18], where Yarmukhamedov completed his construction annotated in [17]. The paper [42] contains a special case of the construction of the Carleman function for the Cauchy problem for the Helmholtz equation in a three-dimensional region of the "cap" type.

Let σ be a positive number. Consider an integer function $K(w)=\exp(\sigma w^2)$ of a complex variable $w\in C$. The contraction of the function K on the vertical line $w=u_0+i\vartheta$ has the form

$$K(u+i\vartheta)=K(u)\exp(2i\sigma uv-\sigma\vartheta^2)$$

i.e., is a decreasing function of . ϑ

Let there be two distinct points $x=(x',x_3)$ and $y=(y',y_3)$ in the space R^3 ; let us assume $r'=|y'-x'|$ and introduce the integral

$$\Phi(x,y)=\frac{-1}{2\pi^2}\frac{1}{K(x_3)}\int_0^\infty \mathrm{Im}\left(\frac{K(w)}{w-x_3}\right)\frac{\cos k\vartheta}{\sqrt{r'^2+\vartheta^2}}dv, \tag{2.4.1}$$

where $w=y_3+i\sqrt{r'^2+\vartheta^2}$, and k is the complex parameter. By isolating the imaginary part,

we get

$$\Phi_\sigma(x,y)=\int_0^\infty p_\sigma(x,y;\vartheta)\cos k\vartheta d\vartheta$$

where $p_\sigma(x,y,\vartheta)$ is given by the relation

$$\frac{-1}{2\pi^2}\frac{e^{\sigma(y_3^2-x_3^2)}e^{-\sigma(r'^2+\vartheta^2)}}{\vartheta^2+r^2}\left((y_3-x_3)\frac{\sin 2\sigma y_3\sqrt{r'^2+\vartheta^2}}{\sqrt{r'^2+\vartheta^2}}-\cos 2\sigma y_3\sqrt{r'^2+\vartheta^2}\right).$$

Thus, the convergence of the nonconvex integral in the right-hand side of the relation (2.4.1) is provided by the multiplier $e^{-\sigma\vartheta^2}$.

Lemma 2.6. *It follows from equality (2.24) that the function* $\Phi(x,y)$ *has the form*

$$\Phi(x,y)=\frac{-1}{4\pi}\frac{e^{-kr}}{r}+R(x,y),$$

(2.4.2)

where $R(x,y)$ *is a twice continuously differentiable function with respect to the variable* $y\in R^3$ *, including the point .* $y=x$

Proof. Let us divide the integration interval in (2.4.1) into two parts, namely, $(0,1]$ and $[1,\infty)$. We will write $I_1(x,y)$ and $I_2(x,y)$ for the first and second integral, respectively. Since $\sqrt{r'^2+\vartheta^2}\geq 1$ for the second integral, $I_2(x,y)$ will be C^∞ -function of the variable $y \in R^n$, including the point . $y = x$

The first integral is transformed as

$$I_1(x,y) = \frac{-1}{2\pi^2}\frac{1}{K(x_3)}\int_0^1 \operatorname{Im}\left(\frac{K(w)-K(x_3)}{w-x_3}\right)\frac{\cos kv}{\sqrt{r'^2+v^2}}dv$$
$$+\frac{-1}{2\pi^2}\int_0^1 \operatorname{Im}(\frac{1}{w-x_3})\frac{\cos kv}{\sqrt{r'^2+v^2}}dv, \tag{2.4.3}$$

where

$$\operatorname{Im}(\frac{1}{w-x_3}) = -\frac{\sqrt{r'^2+\vartheta^2}}{\vartheta^2+r^2}.$$

Since the integer function

$$Q(w) = \frac{K(w)-K(x_3)}{w-x_3}$$

is valid at real w , from the decomposition of

$$\frac{\operatorname{Im} Q(y_3+i\sqrt{r'^2+\vartheta^2})}{\sqrt{r'^2+\vartheta^2}} = \sum_{j=0}^{\infty}(-1)^j\frac{Q^{(j)}(y_3)}{(2j+1)!}(r'^2+\vartheta^2)^j$$

it follows that the first summand in the right-hand side of the relation (2.4.3) is C^∞ -function of the variables x and y . The second summand reduces to the form

$$\int_0^1 \frac{\cos k\vartheta}{\vartheta^2+r^2}dv = \sum_{j=0}^{\infty}(-1)^j\frac{k^{2j}(y_3)}{(2j)!}\int_0^1\frac{\vartheta^{2j}}{\vartheta^2+r^2}dv$$
$$=\int_0^1\frac{1}{\vartheta^2+r^2}dv - \frac{k^2}{2}\int_0^1\frac{\vartheta^2}{\vartheta^2+r^2}dv + \sum_{j=2}^{\infty}(-1)^j\frac{k^{2j}(y_3)}{(2j)!}\int_0^1\frac{\vartheta^{2j}}{\vartheta^2+r^2}d\vartheta$$

to the multiplier $1/2\pi^2$. The first integral in the right-hand side has the form

$$\frac{-\pi}{2}\frac{1}{r} - \int_0^\infty \frac{1}{\vartheta^2+r^2}d\vartheta,$$

and the second integral is

$$-\frac{k^2}{2}\left(1-\int_0^1\frac{r^2}{\vartheta^2+r^2}d\vartheta\right)=-\frac{k^2}{2}+\frac{k^2r^2}{2}\left(-\frac{\pi}{2}\frac{1}{r}-\int_0^\infty\frac{1}{\vartheta^2+r^2}d\vartheta\right).$$

Hence, the second summand in the right part of the relation (2.4.3) is equal to

$$-\frac{\pi}{2}\frac{1}{r}\left(1+\frac{k^2r^2}{2}\right)-\frac{k^2}{2}-\left(1+\frac{k^2r^2}{2}\right)\int_0^\infty\frac{1}{\vartheta^2+r^2}d\vartheta+\sum_{j=2}^\infty(-1)^j\frac{k^{2j}}{(2j)!}\int_0^1\frac{\vartheta^{2j}}{\vartheta^2+r^2}d\vartheta$$

to the multiplier $1/2\pi^2$. Since

$$1+\frac{k^2r^2}{2}=e^{-kr}+kr-\sum_{j=3}^\infty(-1)^j\frac{(kr)^j}{j!}$$

it's easy to see that

$$I_1(x,y)=\frac{-1}{4\pi}\frac{e^{-kr}}{r}+R_1(x,y),$$

where $R_1(x,y)$ is a twice continuously differentiable function with respect to the variable y , including the point $y=x$. Assuming $R=I_1+R_1$, we obtain formula (2.4.2).

Lemma 2.7. *It follows from (2.4.1) that the function* $\Phi(x,y)$ *satisfies the Helmholtz equation* $\Delta\Phi-k^2\Phi=0$ *with respect to* . $y\in R^3\setminus\{x\}$

Proof. Let us denote

$$Q(w)=\frac{-1}{2\pi^2}\frac{1}{K(x_3)}\frac{K(w)}{w-x_3},$$

where $w=y_3+i\sqrt{r'^2+\vartheta^2}$. Then

$$\Delta_y\Phi(x,y)=\int_0^\infty\Delta_y\left(\frac{\operatorname{Im}Q(w)}{\sqrt{r'^2+\vartheta^2}}\right)\cos k\vartheta\, d\vartheta=$$

$$=\frac{1}{2i}\int_0^\infty\Delta_y\left(\frac{Q(w)-Q(\overline{w})}{\sqrt{r'^2+v^2}}\right)\cos k\vartheta\, d\vartheta.$$

The function on which the Laplace operator acts Δ_y , has the form $f(Y,y_n)$, where

$Y=|y'-x'|^2$. An elementary check shows that

$$\Delta_y f(Y,y_n)=4Yf''_{YY}+2(n-1)f'_Y+f''_{y_ny_n}. \tag{2.4.4}$$

Let us denote the partial differentiation operator in the right-hand side of this relation by Δ_{Y,y_n} . Hence, the lemma will be proved if we show that

$$\int_0^\infty \Delta_{Y,y_3}\left(\frac{Q(w)}{\sqrt{Y+\vartheta^2}}\right)\cos k\vartheta\, d\vartheta = k^2\int_0^\infty \frac{Q(w)}{\sqrt{Y+\vartheta^2}}\cos k\vartheta\, d\vartheta,$$
$$\int_0^\infty \Delta_{Y,y_3}\left(\frac{Q(\overline{w})}{\sqrt{Y+\vartheta^2}}\right)\cos k\vartheta\, d\vartheta = k^2\int_0^\infty \frac{Q(\overline{w})}{\sqrt{Y+\vartheta^2}}\cos k\vartheta\, d\vartheta, \tag{2.4.5}$$

at $y\neq x$.

We will mainly prove the first equality in (2.4.5). For this purpose, let us denote the left-hand side of this equality by I . Applying (2.4.4), we obtain

$$I:=\int_0^\infty \Delta_{Y,y_3}\left(\frac{Q(w)}{\sqrt{Y+\vartheta^2}}\right)\cos k\vartheta\, d\vartheta$$
$$=\int_0^\infty\left(4Y\left(\frac{Q(w)}{\sqrt{Y+\vartheta^2}}\right)''_{YY}+4\left(\frac{Q(w)}{\sqrt{Y+\vartheta^2}}\right)'_Y+\left(\frac{Q(w)}{\sqrt{Y+\vartheta^2}}\right)''_{y_3y_3}\right)\cos k\vartheta\, d\vartheta$$

and the simplest calculations show that

$$4\left(\frac{Q(w)}{\sqrt{Y+\vartheta^2}}\right)'_Y=2\left(\frac{iQ'(w)}{Y+\vartheta^2}-\frac{Q(w)}{(Y+\vartheta^2)^{3/2}}\right),$$

$$4Y\left(\frac{Q(w)}{\sqrt{Y+\vartheta^2}}\right)''_{YY}=Y\left(\frac{-Q''(w)}{(Y+\vartheta^2)^{3/2}}-\frac{3iQ'(w)}{(Y+\vartheta^2)^2}+\frac{3Q}{(Y+\vartheta^2)^{5/2}}\right),$$

$$\left(\frac{Q(w)}{\sqrt{Y+\vartheta^2}}\right)''_{y_3y_3}=\frac{Q''(w)}{\sqrt{Y+\vartheta^2}}.$$

Collecting multipliers at the same derivatives of the function Q , we obtain

$$I=I_1+I_2+I_3,$$

where

$$I_1 := \int_0^\infty \frac{\vartheta^2}{(Y+\vartheta^2)^{3/2}} Q''(w)\cos k\vartheta d\vartheta,$$

$$I_2 := \int_0^\infty \frac{2\vartheta^2 - Y}{(Y+\vartheta^2)^2} iQ'(w)\cos k\vartheta d\vartheta,$$

$$I_3 := \int_0^\infty \frac{Y - 2\vartheta^2}{(Y+\vartheta^2)^{5/2}} Q(w)\cos k\vartheta d\vartheta.$$

Since

$$d_v Q'(w) = Q''(w)\frac{i\vartheta}{\sqrt{Y+\vartheta^2}} d\vartheta,$$

then, integrating the first integral by parts, we obtain

$$I_1 = \int_0^\infty \left(\frac{Y-\vartheta^2}{(Y+\vartheta^2)^2}\cos k\vartheta - \frac{k\vartheta}{Y+\vartheta^2}\sin k\vartheta \right) iQ'(w)d\vartheta$$

whence it follows that

$$I_1 + I_2 = \int_0^\infty \frac{\vartheta^2}{(Y+\vartheta^2)^2} iQ'(w)\cos k\vartheta d\vartheta - \int_0^\infty \frac{k\vartheta}{Y+\vartheta^2} iQ'(w)\sin k\vartheta d\vartheta.$$

Here in both integrals in the right-hand side, we again integrate by parts, applying the equality

$$d_v Q(w) = Q'(w)\frac{i\vartheta}{\sqrt{Y+\vartheta^2}} d\vartheta.$$

Then the first integral is transformed to the form

$$-\int_0^\infty \left(\left(\frac{1}{(Y+\vartheta^2)^{3/2}} - \frac{3v^2}{(Y+\vartheta^2)^{5/2}} \right)\cos k\vartheta - \frac{k\vartheta}{(Y+\vartheta^2)^{3/2}} \right) Q(w)d\vartheta,$$

and the second integral is transformed to the form

$$\int_0^\infty \left(-\frac{k\vartheta}{(Y+\vartheta^2)^{3/2}}\sin k\vartheta + \frac{k^2}{\sqrt{Y+\vartheta^2}}\cos k\vartheta \right) Q(w)d\vartheta.$$

Thus,

$$I_1 + I_2 = -\int_0^\infty \frac{Y - 2\vartheta^2}{(Y+\vartheta^2)^{5/2}} Q(w)\cos k\vartheta d\vartheta + k^2 \int_0^\infty \frac{Q(w)}{\sqrt{Y+\vartheta^2}} \cos k\vartheta d\vartheta.$$

From the fact that the first integral in the right-hand side is equal to I_3 , it follows that

$$I = k^2 \int_0^\infty \frac{Q(w)}{\sqrt{Y+\vartheta^2}} \cos k\vartheta d\vartheta,$$

which is exactly what I needed to prove.

Finally, the function is obtained from $Q(w)$ by replacing i by $-i$. Hence,
the calculations do not change and the second equality in (2.4.5) is also proved.
At $k = 0$, the results of this section are still true for the function $K(w) = \exp(\sigma w)$ at $\sigma > 0$, cf. [18], [24], [25],[42].

Carleman's formula

Suppose that X is a bounded region in the upper half-space $x_3 > 0$ of the space R^3 , whose boundary consists of a smooth surface S , lying in the half-space $\{x_3 > 0\}$, and closed pieces of the plane $\{x_3 = 0\}$. Such regions are usually called "cap" type regions. Note that the unit vector of the external normal on the segment $\partial X \setminus S$ is equal to $(0,0-1)$. Consider the problem of finding the electric field E and magnetic field H in X with a given tangential component E_0 in E and normal component H_0 in H at $.S$

Following (2.4.1), let us introduce

$$\Phi_\sigma(x,y) = \frac{-1}{2\pi^2} \frac{1}{K(x_3)} \int_0^\infty \mathrm{Im}\left(\frac{K(w)}{w - x_3}\right) \frac{\cos k\vartheta}{\sqrt{r'^2 + \vartheta^2}} d\vartheta,$$

where , $w = y_3 + i\sqrt{r'^2 + \vartheta^2}$ τ is a complex parameter. Simple calculations show that

$$\Phi_\sigma(x,y) = \int_0^\infty p_\sigma(x,y;\vartheta)\cos k\vartheta d\vartheta,$$

where $p_\sigma(x,y;\vartheta)$ is

$$\frac{-1}{2\pi^2}\frac{e^{\sigma(y_3^2 - x_3^2)}e^{-\sigma(r'^2+\vartheta^2)}}{\vartheta^2 + r^2}\left((y_3 - x_3)\frac{\sin 2\sigma y_3\sqrt{r'^2+\vartheta^2}}{\sqrt{r'^2+\vartheta^2}} - \cos 2\sigma y_3\sqrt{r'^2+\vartheta^2}\right).$$

Hence, we obtain

$$p_\sigma = \frac{1}{2\pi^2}\frac{e^{-\sigma x_3^2}e^{-\sigma(r'^2+\vartheta^2)}}{\vartheta^2 + r^2},$$

$$\partial y_3 p_\sigma = \frac{1}{\pi^2}\frac{e^{-\sigma x_3^2}e^{-\sigma(r'^2+\vartheta^2)}}{(\vartheta^2 + r^2)^2}x_3\left(1 + \sigma(\vartheta^2 + r^2)\right)$$

on the plane $y_3 = 0$. Substituting $\Phi_\sigma(x,y)$ into the formula for the fundamental

solution $\Phi(x-y)$ for the Maxwell operator M we conclude that $\Psi_\sigma(x,y)$ is the Carleman function of the Cauchy problem in the domain X with data at S , parameterized with the parameter $.\sigma$

$$\Psi_\sigma(x,y) = \begin{pmatrix} (1/ik)d^*d & -d^* \\ -d & -(1/ik)dd^* \end{pmatrix}\Phi_\sigma(x,y)$$

Lemma 2.8. *Matrix-valued function* $\Psi_\sigma(x,y)$ *on sets* $R^3 \times (R^3 \setminus \{y = x\})$, *satisfying*

$$M(\partial_x)\Psi_\sigma(x,y) = 0,$$

$$M'(\partial_y)(\Psi_\sigma(x,y))^T = 0.$$

Proof. The proof follows from the properties of the fundamental solution $\Phi(x-y)$ of the Maxwell operator.

Theorem 2.6. *Let* (E,H) *be an electromagnetic wave in* X *, continuous up to* S *. Then the limit equality*

$$\begin{pmatrix} E(x) \\ H(x) \end{pmatrix} = \lim_{\sigma\to\infty} \int_S \Psi_\sigma(x,y) \begin{pmatrix} in(H) \\ -\nu \wedge t(E) \end{pmatrix} ds$$

is performed uniformly on every compact subset of. X

Proof. As we know, the value of the regular solution of Maxwell's equations at the point x inside X is expressed through the values $t(E)$ and $n(H)$ at the boundary X . Using the Stratton-Chu formula from Theorem 2.5

$$\begin{pmatrix} E(x) \\ H(x) \end{pmatrix} = \int_S \Psi_\sigma(x,y) \begin{pmatrix} in(H) \\ -\nu \wedge t(E) \end{pmatrix} ds + \int_{\partial X \setminus S} \Psi_\sigma(x,y) \begin{pmatrix} in(H) \\ -\nu \wedge t(E) \end{pmatrix} ds$$

Thus, the convergence of the integral in the right-hand side of the relation $\Phi_\sigma(x,y)$ is ensured by the multiplier $e^{-\sigma\vartheta^2}$. If $\sigma \to \infty$ the expression , $\Psi_\sigma(x,y)$ $p_\sigma(x,y;\vartheta)$ and its partial derivative $\partial_{y_3} p_\sigma(x,y;\vartheta)$ tend to zero exponentially in the plane $\{x_3 > 0\}$. It follows that the part of the boundary integral over $\partial X \setminus S$ tends to zero at $\{x_3 = 0\}$. This establishes the formula sought.

CHAPTER III. THE CAUCHY PROBLEM FOR AN ELLIPTIC SYSTEM OF MAXWELL-TYPE EQUATIONS IN " n " DIMENSIONAL SPACE

§3.1 Maxwell's abstract equations

Regarding the presentation of the mathematical treatment of the theory of electromagnetic scattering, we refer the reader to [21, 24], etc.

The system of equations (2.12) can be generalized to arbitrary complexes of differential operators on C^{∞} defined on finite and infinite manifolds X . More precisely, let us consider the complex of first order differential operators on X , acting on sections of vector separations over X , i.e.

$$0 \longrightarrow C^{\infty}(X,F^{0}) \xrightarrow{A^{0}} C^{\infty}(X,F^{1}) \xrightarrow{A^{1}} \ldots \xrightarrow{A^{N-1}} C^{\infty}(X,F^{N}) \longrightarrow 0 \quad , \tag{3.1.1}$$

where $A^{i} \in Diff^{1}(F^{i},F^{i+1})$ – is the vector space of all differential operators of order 1 mapping $C^{\infty}(X,\mathrm{F}^{i})$ to $C^{\infty}(X,\mathrm{F}^{i+1})$ with condition $A^{i+1}A^{i}=0$ for all i . As usual we write instead of $A^{i}u$ simply Au for $u \in C^{\infty}(X,\mathrm{F}^{i})$, when no confusion can arise. Here F^{i} is a smooth vector stratification of rank k_{i} which is different from zero at $i=0,1,...,N$. We define on F^{i} an Hermite metric, i.e., the scalar products $x \mapsto (v,w)_{x}$ in the layers of the stratification F_{x}^{i} , which depend smoothly on the point $x \in X$. This defines a conjugate linear isomorphism ★ of F^{i} to an algebraically dual stratification F^{i*} with $\bigstar\langle w,v\rangle_{X}=(v,w)_{X}$ for v , w . $\in F_{X}^{i}$

Fix a smooth positive volume form of dx to X . This gives the scalar product of $C_{comp}^{\infty}(X,\mathrm{F}^{i})$

$$(u,v)=\int_X (u(x),v(x))_X dx$$

for $u,v \in C^{\infty}_{comp}(X,F^{i})$. Let us denote by $L^2(X,F^{i})$ the replenishment of the space by the corresponding norm. Moreover, we introduce a formally conjugate operator $A^{i*} \in Diff^{1}(F^{i+1},F^{i})$ for each operator A^{i} , requiring $(Au,g)=(u,A^{*}g)$ for all $u \in C^{\infty}(\Omega,F^{i})$ and $g \in C^{\infty}(\Omega,F^{i+1})$ whose carriers do not intersect on the boundary . X

Formally, the self-adjoint operators $\Delta^{i} = A^{i*}A^{i} + A^{i-1}A^{i-1*}$ on X are called the Laplaceans of the complex (3.1.1). The ellipticity of this complex follows from the fact that each Laplace operator $\Delta^{i} \in Diff^{2}(F^{i})$ is a second order elliptic operator (see[37])

For any $-1 \le i \le N$, the system of Maxwell's equations in complexes (2.4.3) at step i is written

$$\begin{aligned} -\varepsilon(d/dt)u + A^{*}f &= \sigma u, \\ \mu(d/dt)f + Au &= 0, \end{aligned}$$

where u and f are unknown functions from t with values in the stratifications F^{i} and F^{i+1} , respectively. When searching for harmonic solutions of these equations in the form of

$$\begin{aligned} u(t,x) &= (\varepsilon + i\sigma/\omega)^{-1/2} e^{-i\omega t} u(x), \\ f(t,x) &= \mu^{-1/2} e^{-i\omega t} f(x), \end{aligned}$$

fit the stationary equations for u and f , which depend only on points of the base manifold . X

These

$$\begin{aligned} iku + A^{*}f &= 0, \\ -ikf + Au &= 0, \end{aligned}$$

where k is the wave constant defined above.

Definition 3.1. *Let* $-1 \leq \mathrm{i} \leq n$. *We define the Maxwell operator for the complex (3.1.1) at step* i *as follows:*

$$M^{\mathrm{i}} = \begin{pmatrix} ik & A^{\mathrm{i}*} \\ A^{\mathrm{i}} & -ik \end{pmatrix} .$$

As is customary in homological algebra, we omit the index i in M^{i} this as seen from the context.

By definition, M^{i} is a first order differential operator from sections of a stratification $F^{\mathrm{i}} \oplus F^{\mathrm{i}+1}$ into sections of the same stratification over X . This operator cannot be a 1st order elliptic operator in the classical sense at $N=2$. On the other hand applying A^{*} to both parts of the equality $iku + A^{*}f = 0$, we conclude that $A^{\mathrm{i}\text{-}1*}u = 0$ excluding $k=0$. Similarly from $-ikf + Au = 0$ it follows that $A^{\mathrm{i}+1}f = 0$, excluding $k=0$. Complementing Maxwell's equations with their differential consequences $A^{\mathrm{i}\text{-}1*}u = 0$ and $A^{\mathrm{i}+1}f = 0$, we arrive at a system of first order differential equations for u and f whose classical symbols are injective. There is another way to indicate that there exists a differential operator C^{i} from sections of a stratification $F^{\mathrm{i}} \oplus F^{\mathrm{i}+1}$ into sections of the same stratification, such that $C^{\mathrm{i}}M^{\mathrm{i}}$ are a second order differential operator on X an elliptic in the classical sense. An easy calculation shows that

$$C^{\mathrm{i}} = \begin{pmatrix} ik + (1/ik)A^{\mathrm{i}\text{-}1}A^{\mathrm{i}\text{-}1*} & A^{\mathrm{i}*} \\ A^{\mathrm{i}} & -ik - (1/ik)A^{\mathrm{i}+1*}A^{\mathrm{i}+1} \end{pmatrix} \qquad (3.1.2)$$

Lemma 3.1. *As defined above,* C^{i} satisfies

$$C^{\mathrm{i}}M^{\mathrm{i}} = M^{\mathrm{i}}C^{\mathrm{i}} = \begin{pmatrix} \Delta^{\mathrm{i}} - k^2 & 0 \\ 0 & \Delta^{\mathrm{i}+1} - k^2 \end{pmatrix}.$$

Proof.

$$\begin{pmatrix} ik+(1/ik)A^{i-1}A^{i-1*} & A^{i*} \\ A^{i} & -ik-(1/ik)A^{i+1}A^{i+1*} \end{pmatrix}\begin{pmatrix} ik & A^{i*} \\ A^{i} & -ik \end{pmatrix}=$$

$$=\begin{pmatrix} -k^2+A^{i-1}A^{i-1*}+A^{i*}A^{i} & (ik+(1/ik)A^{i-1}A^{i-1*})A^{i*}-ikA^{i*} \\ ikA^{i}+(-ik-(1/ik)A^{i+1*}A^{i+1})A^{i} & A^{i}A^{i*}+A^{i+1*}A^{i+1}-k^2 \end{pmatrix}=$$

$$\begin{pmatrix} \Delta^{i}-k^2 & ikA^{i*}+(1/ik)A^{i-1}A^{i-1*}A^{i*}-ikA^{i*} \\ ikA^{i}-ikA^{i}-(1/ik)A^{i+1*}A^{i+1}A^{i} & \Delta^{i+1}-k^2 \end{pmatrix}$$

$$=\begin{pmatrix} \Delta^{i}-k^2 & 0 \\ 0 & \Delta^{i+1}-k^2 \end{pmatrix}.$$

§3.2 The Stratton-Chu formula

For the classical Maxwell equations, the so-called Stratton-Chu formula [36] underlies the analytical methods used, [21]. From the point of view of modern analysis this formula is a very special case of the Green's formula, see [37, 2.5.4].

Let X be a compact manifold with edge that is smoothly embedded in a larger C^{∞} manifold X', and (3.1.1) is defined over the whole X'. If $X'=R^{n}$, , then the image for X can be any bounded region in the R^{n}

Suppose that $\Delta^{i}-k^2$ and $\Delta^{i+1}-k^2$ are second-order elliptic operators and satisfy the condition of uniqueness in the small Cauchy problem on X'. Then [29] these operators have left fundamental solutions on X', which we denote by G^{i} and G^{i+1}, respectively. [37, p 105],[32] and [9].

These latter are classical pseudodifferential operators of order - 2 acting on sections of vector stratifications F^{i} and F^{i+1} over . X'

Lemma 3.2. *Pseudodifferential operator*

$$\Phi^{i} = \begin{pmatrix} G^{i}(ik + (1/ik)A^{i-1}A^{i-1*}) & G^{i}A^{i*} \\ G^{i+1}A^{i} & G^{i+1}(-ik - (1/ik)A^{i+1*}A^{i+1}) \end{pmatrix}$$

there is a left fundamental solution of the Maxwell operator M^{i} *to* . X'

Proof. It follows immediately from lemma 1.1 that

$$\Phi^{i} = \begin{pmatrix} G^{i} & 0 \\ 0 & G^{i+1} \end{pmatrix} \circ C^{i}$$

is the left fundamental solution of M^{i} . It remains to substitute the explicit expression (3.1.2) for . C^{i}

Note that Φ^{i} is a pseudodifferential operator of order 0 (not - 1) on sections of the vector stratification $F^{i} \oplus F^{i+1}$ over X' . This corresponds to the fact that the Maxwell operator is elliptic in the sense of Douglas Nirenberg with $s_{i} = t_{i} = 0$, if $k \neq 0$. As usual, we use the same letter to denote the operator Φ^{i} of a vector stratification $F^{i} \oplus F^{i+1}$ over the distribution sections of the same stratification and its Schwarz kernel over $X' \times X'$ which is denoted by $\Phi^{i}(x, y)$.

Theorem 3.1. *For every* $u \in H^{1}(X, F^{i})$ *and* $f \in H^{1}(X, F^{i+1})$ *satisfying* $M^{i}(u, f) = 0$ *inside* X *it follows that*

$$\int_{\partial X} \Phi^{i}(x, y) \begin{pmatrix} \sigma(A^{i})^{*}(y, i\nu(y)) f(y) \\ \sigma(A^{i})(y, i\nu(y)) u(y) \end{pmatrix} ds(y) = \begin{cases} \begin{pmatrix} u(x) \\ f(x) \end{pmatrix}, & \text{если} \quad x \in X \setminus \partial X, \\ 0, & \text{если} \quad x \in X' \setminus X. \end{cases}$$

Here, $\nu(y)$ is the outer unit vector normal to the boundary at the point $y \in \partial X$, ds the surface measure at X , and $\sigma(A^{i})$ the (classical) symbol for . A^{i}

Proof. This formula is a special case of the general Green's formula (see [37, 2.5.4]) for

$$\sigma(M^{i}) = \begin{pmatrix} 0 & \sigma(A^{i})^{*} \\ \sigma(A^{i}) & 0 \end{pmatrix}$$

Let us write $t(u)$ and $n(f)$ for the tangential part u and the normal part f at the boundary X with respect to the complex (3.1.1), respectively (see [40]). Then we obtain

$$\begin{pmatrix} \sigma(A^{i})^{*}(y, i\upsilon) f \\ \sigma(A^{i})(y, i\upsilon) u \end{pmatrix} = \begin{pmatrix} i\sigma(\Delta^{i})^{*}(y,\upsilon) n(f) \\ i\sigma(A^{i})(y,\upsilon) t(u) \end{pmatrix}$$

(3.2.1)

at ∂X . The same reasoning shows that Theorem 3.1 still holds for all solutions (u, f) of Maxwell's equations inside X , such that $t(u)$ and $n(f)$ have weak limits at the boundary ∂X , (see [39; 9.4]).

For the classical Maxwell equations, the formula of Theorem 1.2 is known as the Stratton-Chu formula, (see [36]). In this case it has a better structure, the fundamental solution Φ^{i} can be written down explicitly. This can indeed be done in the context of the so-called Dirac-type complex in R^{n} , which is characterized by the property that all Laplaceans of Δ^{i} are diagonal operators, with the usual (non-negative) scalar Laplace operator on the diagonal. In other words, $\Delta^{i} = -E_{k_i}\Delta$ is satisfied for all i from 0 to N , where E_{k_i} is a unit matrix of order $k_i \times k_i$ and $\Delta = \partial_1^2 + ... + \partial_n^2$. Let $e(x)$ be a standard fundamental solution of the convolution type of the Helmholtz operator $\Delta + k^2$ in R^n, which goes to zero at ∞. If $n = 3$, then

$$e(x) = -\frac{1}{4\pi}\frac{\exp(ik\,|\,x\,|)}{|\,x\,|}$$

outside the origin.

Corollary 3.1. *Let (3.1.1) complex consist of differential operators with constant coefficients of Dirac type and* X *a bounded closed region with smooth boundary in* R^n. *If* $u \in H^1(X, F^i)$ *and* $f \in H^1(X, F^{i+1})$ *satisfies* $M^i(u, f) = 0$ *inside* X, *then*

$$\begin{pmatrix} (1/ik)A^{i*}A^i & -A^{i*} \\ -A^i & (1/ik)A^iA^{i*} \end{pmatrix} \int_{\partial\Omega} e(x-y) \begin{pmatrix} in(f) \\ i\tau(A^i)(\upsilon)t(u) \end{pmatrix} dS = \begin{pmatrix} u(x) \\ f(x) \end{pmatrix} \qquad (3.2.2)$$

for all $x \in X \setminus \partial X$ *and the left side disappears outside*. X

The formula (3.2.2) extends modulo smoothing operators to the general Maxwell equations.

Proof. For a Dirac-type complex we can choose $G^i u = -e * u$ the partition distribution and functions F^i with a compact carrier in R^n, where we denote by $e * u$ the convolutions e and u. The Schwarz kernel of G^i is obviously $E_{k_i} e(x-y)$. Thus, the Schwarz kernel Φ^i is

$$-C^{i/}(y, \partial_y)(E_{k_i + k_{i+1}} e(x-y)) = -C^i(y, \partial_x) e(x-y),$$

variable y in C^i can be neglected if the coefficients are constant. From (3.2.2) it is easy to deduce that

$$C^i = \begin{pmatrix} (1/ik)(\Delta^i - k^2 - A^{i*}A^i) & A^{i*} \\ A^i & -(1/ik)(\Delta^{i+1} - k^2 - A^iA^{i*}) \end{pmatrix},$$

and also

$$\Phi^{\mathrm{i}}(x,y)=-\begin{pmatrix}-(1/ik)A^{\mathrm{i}*}A^{\mathrm{i}}e(x-y) & A^{\mathrm{i}*}e(x-y)\\ A^{\mathrm{i}}e(x-y) & (1/ik)A^{\mathrm{i}}A^{\mathrm{i}*}e(x-y)\end{pmatrix}.$$

At $R^n\setminus\{y\}$, for $(\Delta+k^2)e=0$ far from the origin. To complete the proof it is enough to use Theorem 1 and take into account that $\sigma(\Delta^i)(\upsilon)=|\upsilon|^2 E_{k_i}=E_{k_i}$ in our particular case.

As is easy to verify,

$$M^{\mathrm{i}}\begin{pmatrix}\frac{1}{ik}A^{\mathrm{i}*}A^{\mathrm{i}} & -A^{\mathrm{i}*}\\ -A^{\mathrm{i}} & -\frac{1}{ik}A^{\mathrm{i}}A^{\mathrm{i}*}\end{pmatrix}=\begin{pmatrix}0 & -\frac{1}{ik}A^{\mathrm{i}*}(\Delta^{\mathrm{i}+1}-k^2)\\ \frac{1}{ik}A^{\mathrm{i}}(\Delta^{\mathrm{i}}-k^2) & 0\end{pmatrix}$$

hence the left-hand side of (3.2.2) satisfies the Maxwell equations in $R^n\setminus\partial X$ for all integrable functions $t(u)$ and $n(f)$ on the boundary. Thus, we arrive at what is usually called the Cauchy integral associated with Maxwell's equations (see [37, 3.2.3])

§3.3. Criterion of solvability of the Cauchy problem

Combining the systems of Maxwell's equations with the data from $t(u)$ and $n(f)$ we determine u and f at the boundary, for the

$$n(u)=-(1/ik)n(A^*f),$$
$$t(f)=(1/ik)t(Au)$$

$n(A^*f)$ and $t(Au)$ is uniquely determined by $n(f)$ and $t(u)$ (the so-called tangent operators on ∂X , see[39, 3.1.5]). Thus, the Cauchy problem for the system of Maxwell's equations X consists in finding the solution of (u,f) in $M^{\mathrm{i}}(u,f)=0$ with data $t(u)$ and $n(f)$ on a part of the surface ∂X [13]. To obtain constructive results, we restrict ourselves to complexes of Dirac-type differential operators with constant coefficients in R^n , as defined in §3.2.

Let S be the open part of the boundary of the set X in the relative topology. The Cauchy problem for solutions of the system of Maxwell's equations in X with data at S is as follows. Given the cross sections u_0 and f_0 of the stratification F_t^{i} over S , find a solution (u,f) satisfying $M\ (u,f)=0$ inside X such that $t(u)=u_0$ and $n(f)=f_0$ on . S

This is a bit cumbersome, since the behavior of the solution near the boundary S requires careful study. We restrict ourselves to considering the case where u_0 and f_0 are continuous sections of the stratification F^{i} . If it exists, then the solution (u,f) must be continuous up to the interior of S , and we avoid discussing the weak boundary values of (u,f) at S . Thus, consider the problem

$$\begin{cases} M^{\mathrm{i}}(u,f)=0 \text{ в } & X\setminus\partial X, \\ t(u)=u_0 \quad \text{на} & S, \\ n(f)=f_0 \quad \text{на} & S. \end{cases}$$

(3.3.1)

It is well known that this problem has at most one solution in any reasonable function space in . X

To study problem (3.3.1), we conduct the integral completely defined by the Cauchy data u_0 and f_0 on S , namely

$$G(u_0,f_0)(x)=\int_S \Phi^{\mathrm{i}}(x-y)\begin{pmatrix} if_0 \\ i\sigma(A^{\mathrm{i}})(\nu)u_0 \end{pmatrix} ds$$

for $x\in R^n\setminus\overline{S}$. Since the fundamental solution of $\Phi^{\mathrm{i}}(x-y)$ is real analytic off-diagonal to $R^n\times R^n$ it follows that $G(u_0,f_0)$ is real analytic in addition to . $\overline{S}$

Moreover, the potential $G(u_0, f_0)$ satisfies Maxwell's equations $M^{i}(u,f)=0$ in $R^n \setminus \overline{S}$. In particular, the components of the vector-function $G(u_0, f_0)$ turn out to be solutions of the (scalar) Helmholtz equation $(\Delta+k^2)G=0$ in the complement of $\overline{S}$.

When x passes over the hypersurface S, the integral $G(u_0, f_0)(x)$ has a jump. The corresponding transition formulas are very similar to the classical Sohotsky-Plemel formulas, see [9]. [9].

Theorem 3.2. *For there to exist a solution* (u, f) *of the Cauchy problem (3.33) continuous up to* S, *it is necessary and sufficient that the integral* $G(u_0, f_0)$ *can be continued from* $R^n \setminus X$ *through* S *to* X *as a real analytic function.*

Proof ***Necessity***. Suppose that there exists a solution (u, f) of the Cauchy problem (3.3.1), continuous up to S. Let us determine the cross sections U and F from F^{i} and F^{i+1} to $R^n \setminus \partial X$ by the formula

$$\begin{pmatrix} U(x) \\ F(x) \end{pmatrix} = \begin{cases} G(x_0, y_0) - \begin{pmatrix} u \\ f \end{pmatrix} & в \;\; X \setminus \partial X, \\ G(u_0, f_0) \quad в \quad R^n \setminus X. \end{cases}$$

Let us write $U^{\pm}$ to contract U into open sets $X \setminus \partial X$ and $R^n \setminus \overline{X}$, respectively, and similarly for F. We can assume, by reducing X, if necessary, that u and f are continuous up to the boundary of X. Using Theorem 3.1, (paragraph 3.2) we obtain

$$\begin{pmatrix} U^{+}(x) \\ F^{+}(x) \end{pmatrix} = -\int_{\partial X \setminus S} \Phi^{i}(x-y) \begin{pmatrix} in(f) \\ i\sigma(A^{i})(\nu)t(u) \end{pmatrix} ds$$

for all x inside X. It follows that U^+ and F^+ continue through S as real analytic sections of F^{i} and F^{i+1} on all $R^n \setminus \partial X \cup S$. Moreover, applying Theorem 3.1 once again we have

$$-\int_{\partial X \setminus S} \Phi^{i}(x,y)\begin{pmatrix} in(f) \\ \imath\sigma(A^{i})(\nu)t(u)\end{pmatrix} ds =$$

$$= \int_{S} \Phi^{i}(x,y)\begin{pmatrix} in(f) \\ i\sigma(A^{i})(\nu)t(u)\end{pmatrix} ds = \begin{pmatrix} U^-(x) \\ F^-(x)\end{pmatrix}$$

for $x \in R^n \setminus X$. Hence, $G(u_0, f_0)$ continues in $R^n \setminus X$ through S inside X as a real analytic function, which was required to prove.

***Sufficiency*.** Inversely, let U and F be real analytic sections of the stratifications F^{i} and F^{i+1} over $(R^n \setminus \partial X) \cup S$ such that

$$\begin{pmatrix} U(x) \\ F(x)\end{pmatrix} = G(u_0, f_0)$$

away from X .Then $M^{i}(U,F) = 0$ is in $R^n \setminus X$. Since $M^{i}(U,F)$ is real analytic in $(R^n \setminus \partial X) \cup S$, it actually also converges to zero in . X

Let's

$$\begin{pmatrix} u(x) \\ f(x)\end{pmatrix} = G(u_0, f_0)(x) - \begin{pmatrix} U(x) \\ F(x)\end{pmatrix}$$

for x inside the neighborhood of X. From what has already been proved that u and f are continuous up to S and satisfy Maxwell's equations. Let us show that (u, f) is the desired solution of (3.3.1). To verify this, it remains to check that $t(u) = u_0$ and $n(f) = f_0$ are at S. To do this, let us denote by $G_e(u_0, f_0)$ and $G_m(u_0, f_0)$ the components of $G(u_0, f_0)$, which are the electric and magnetic fields, respectively. Since U and F are continuous in $(R^n \setminus \partial X) \cup S$, by the Sohotsky-Plemel formula we have

$$t(u) = t(G_e(u_0, f_0)^+ - U^+) =$$
$$= t(G_e(u_0, f_0)^+ - U^-) =$$
$$= t(G_e(u_0, f_0)^+ - G_e(u_0, f_0)^-) =$$
$$= u_0$$

at S . Similarly,

$$n(f) = n(G_m(u_0, f_0)^+ - F^+) =$$
$$= n(G_m(u_0, f_0)^+ - F^-) =$$
$$= n(G_m(u_0, f_0)^+ - G_m(u_0, f_0)^-) =$$
$$= f_0$$

at S , which is our assertion.

Theorem 3.2 can be generalized by saying that any explicit formula for the analytic continuation from $R^n \setminus X$ through S to the interior of X leads to a formula for solutions to the Cauchy problem with data on S . This idea goes back at least as far as [3], [19].

§3.4 Decomposition of the fundamental solution of the Helmholtz equations.

In this paragraph, we extend the results of Section 6 of [30],[34] from the case of $n = 3$ to the case of arbitrary $. n > 3$

We apply the double orthogonality basis method to the Cauchy problem (3.3.1) in the special case where X is part of a ball $B = B(0, R)$ centered at the origin and radius $R > 0$. Let S be a smooth closed hypersurface in B , does not contain the point $x = 0$ and splits B into two regions. Let us denote by X the closed region not containing the origin of coordinates. This boundary ∂X consists of S and a part of the sphere ∂B in R^n , cf. Figure 1. The advantage of using the region X of the above is that the problem is reduced to an analytic continuation from a small sphere around 0 to . B

By virtue of the above, the double layer potential $G(u_0, f_0)$ satisfies both $M^i G(u_0, f_0) = 0$, and $(\Delta + k^2)G(u_0, f_0) = 0$ outside the closure S . The latter equation is a scalar Helmholtz equation and follows from the previous one. The idea is to use bases with dual orthogonality in the Hilbert spaces of solutions of the Helmholtz equation to derive analytic continuation conditions from $B(0,r)$ to $B(0,R)$ for . $G(u_0, f_0)$

In spherical coordinates the Helmholtz operator $\Delta + k^2$ in the space R^n takes the form

$$\Delta + k^2 = \frac{1}{r^2}\left(\left(r\frac{\partial}{\partial r}\right)^2 + (n-2)r\frac{\partial}{\partial r} + k^2 r^2 - \Delta_{S^{n-1}}\right), \tag{3.4.1}$$

where $\Delta_{S^{n-1}}$ is the Laplace-Beltrami operator on the unit sphere. Recall that k is an arbitrary complex number with . $\operatorname{Im} k \geq 0$

We can use the Fourier method of separation of variables to solve the homogeneous equation $(\Delta + k^2)u = 0$. Exactly, writing $u(r,w) = g(r,k)h(w)$, we get two separate equations for the functions g and h ,

$$\left(\left(r\frac{\partial}{\partial r}\right)^2+(n-2)r\frac{\partial}{\partial r}+k^2r^2\right)g=cg,$$

$$\Delta_{S^{n-1}}h=ch,$$

with an arbitrary constant c .

The second equation has nonzero solutions if and only if c is an eigenvalue of the operator $\Delta_{S^{n-1}}$. These are well known $c=\nu(\nu+n-2)$, for $\nu=0,1,...,$ see. [39] and elsewhere. The corresponding eigenfunctions $\Delta_{S^{n-1}}$ are spherical harmonics $h_\nu(w)$ of degree ν , i.e.

$$\Delta_{S^{n-1}}h_\nu=\nu(\nu+n-2)h_\nu \qquad . (3.4.2)$$

Let us now consider an ordinary differential equation on the variable . $r>0$

$$\left(\left(r\frac{\partial}{\partial r}\right)^2+(n-2)r\frac{\partial}{\partial r}+(k^2r^2-\nu(\nu+n-2))\right)g(r,k)=0.$$

(3.4.3)

These are Bessel equations, and the space of its solutions is two-dimensional. For example, if $n=3$ and $k=0$, then $g(r,0)=ar^\nu+br^{-\nu-1}$ with arbitrary constants a and b is a general solution of (3.4.3). In this situation, any function $r^\nu h_\nu(\varphi)$ is a homogeneous harmonic polynomial. In the general case, the solution space of (3.4.3) contains a one-dimensional subspace of functions bounded at the point $r=0$, see [40]. [40].

At $\nu=0,1,...,$ fixed is a nonzero solution $g_\nu(r,k)$ of equation (3.4.3), which is bounded at $r=0$. Then

$$(\Delta+k^2)(g_\nu(r,k)h_\nu(w)=0 \qquad (3.4.4)$$

on all R^n . Indeed, by virtue of (3.4.1), (3.4.2) and (3.4.3) we conclude that this equality holds in $R^n \setminus \{0\}$. We now use the fact that $(g_\nu(r,k)h_\nu(w)$ is bounded at the origin to make sure that (3.4.4) is satisfied.

As it is known that the number $J(\nu)$ of linearly independent spherical harmonics of degree ν , is equal to

$$J(\nu)=\frac{(2\nu+n-2)(\nu+n-3)!}{(n-2)!\nu!}\quad .$$

Let us construct the orthonormalized basis

$$\{h_\nu^j\}_{\substack{\nu=0,1,\dots\\ j=1,\dots,J(\nu)}}$$

at . $L^2(S^{n-1})$

Lemma 3.3. *For any* $R>0$ *system*

$$\{b_\nu^{(j)}(r,w,k):=g_\nu(r,k)h_\nu^{(j)}(w)\}_{\substack{\nu=0,1,\dots\\ j=1,\dots,J(\nu)}}$$

(3.4.5)

is an orthogonal basis in the subspace $L^2(B(0,R))$ *, consisting of solutions of the Helmholtz equation* . $(\Delta+k^2)u=0$

Proof. Indeed, $\{h_\nu^{(j)}\}$ is an orthonormalized basis in the space $L^2(S^{n-1})$ on the unit sphere, the system (3.4.5) is orthogonal in $L^2(B(0,R))$, because

$$\left(b_\mu^{(i)},b_\nu^{(j)}\right)_{L^2(B(0,R))}=\left(h_\mu^{(i)},h_\nu^{(j)}\right)_{L^2(S^{n-1})}\int_0^R g_\mu(r,k)\overline{g_\nu(r,k)}\cdot r^{n-1}dr=0$$

for $\mu\neq\nu$ or $i\neq j$. Finally, since the harmonic system $\{h_\nu^{(j)}\}$ is dense in $C^\infty(S^{n-1})$ we see that the system (3.38) is dense in the subspace $L^2(B(0,R))$

, consisting of solutions of the Helmholtz equation in the ball, which completes the proof.

For any fixed $y \in R^n$ with $|y| > R$, the fundamental solution $e(x-y)$ of the Helmholtz equation satisfies $(\Delta + k^2)e(x-y) = 0$ in $B(0,R)$ and obviously it is quadratically integrable in the ball. Therefore, $e(x-y)$ can be represented in $B(0,R)$ by its Fourier series over the orthogonal system (3.4.5)

$$e(x-y) = \sum_{\nu=0}^{\infty} \sum_{j=1}^{J(\nu)} c_\nu^{(j)}(y,k) b_\nu^{(j)}(x,k) \qquad , (3.4.6)$$

series converges in $L^2(B(0,R))$ - by norms. The Fourier coefficients of $c_\nu^{(j)}(y,k)$ are determined by well-known formulas via the scalar product in $L^2(B(0,R))$. These formulas show that $c_\nu^{(j)}(y,k)$ are real analytic functions from y subject to $|y| > R$. Moreover, they satisfy the Helmholtz equation in the complement of $B(0,R)$. On the other hand, the system (3.4.5) is orthogonal to any space $L^2(B(0,R))$ with $R > 0$. Hence, the coefficients of $c_\nu^{(j)}(y,k)$ are effectively independent of the particular choice of R satisfying the condition $R < |y|$. This shows that $c_\nu^{(j)}(y,k)$ are solutions of the Helmholtz equation from the origin in R^n. Since $e(x-y) = O(|x-y|^{2-n})$ at $x \to y$, we can bias $R \to |y|$ in the formulas for $c_\nu^{(j)}(y,k)$. This immediately gives the equations

$$c_\nu^{(j)}(y,k) = \left(e(x-y), b_\nu^{(j)}(x,k)\right)_{L^2(B(0,|y|))} / \int_0^{|y|} |g_\nu(r,k)|^2 r^{n-1} dr$$

for $\nu = 0,1,\ldots$ and . $j = 1,\ldots,J(\nu)$

Lemma 3.4. *In the cone* $\{(x,y)\in R^n\times R^n :| x|/| y|<1\}$ *, has a Fourier series expansion (3.4.6), where the series converges uniformly together with all its derivatives on compact subsets of the cone.*

Proof. The decomposition into Fourier series is a direct consequence of Lemma 3.3. On uniform convergence on compact subsets of the cone presents a more subtle problem. It can be treated in a familiar way since $b_\nu^{(j)}$ are solutions of the Helmholtz equation.

§3.5. Carleman's formula for the system of equations Maxwellian

Substituting (3.4.6) into the formula for the fundamental solution of $\Phi^{\mathrm{i}}(x-y)$ for M^{i} , we obtain

$$\Phi^{\mathrm{i}}(x-y)=\sum_{\nu=0}^{\infty}\Phi_\nu^{\mathrm{i}}(x,y) \tag{3.5.1}$$

where the series converges uniformly together with all its derivatives on compact subsets of the cone $\{(x,y)\in R^n\times R^n :| x|/| y|<1\}$ and

$$\Phi_\nu^{\mathrm{i}}(x,y)=\begin{pmatrix}(1/ik)A^{\mathrm{i}*}(\partial_x)A^{\mathrm{i}}(\partial) & -A^{\mathrm{i}*}(\partial_x)\\ -A^{\mathrm{i}}(\partial_x) & -(1/ik)A^{\mathrm{i}}(\partial_x)A^{\mathrm{i}*}(\partial_x)\end{pmatrix}\sum_{j=1}^{J(\nu)}c_\nu^{(j)}(y,k)b_\nu^{(j)}(x,k)$$

for $\nu=0,1,\ldots$. We get the same formula if we substitute together $-\partial_y$ for the derivatives of . ∂_x

Lemma 3.5. *Each member of* $\Phi_\nu(x,y)$ *is a real-analytic matrix-valued function on the set* $R^n\times(R^n\setminus\{0\})$ *, satisfying*

$$M^{\mathrm{i}}(\partial_x)\Phi_\nu^{\mathrm{i}}(x,y)=0,$$
$$M^{\mathrm{i}'}(\partial_y)(\Phi_\nu^{\mathrm{i}}(x,y))^T=0.$$

Proof. These properties are obvious by the very construction of degenerate kernels $\Phi_\nu^{\mathrm{i}}(x,y)$. The peculiarity at $y=0$ is due to the assumption of

$$\int_0^{|y|} |g_\nu(r,k)|^2 r^{n-1} dr.$$

A series expansion similar to (3.4.6) with terms Φ_ν^{i} satisfying the transposed equation $M^{\mathrm{i}'}(\partial_y)(\Phi_\nu^{\mathrm{i}}(x,y))^T = 0$ is already enough to obtain an explicit formula for solutions of the Cauchy problem (3.3.1).

Let's

$$R_N^{\mathrm{i}}(x,y) = \Phi^{\mathrm{i}}(x-y) - \sum_{\nu=0}^{N} \Phi_\nu^{\mathrm{i}}(x,y)$$

for $(x,y) \in R^n \times (R^n \setminus \{0\})$. Thus, we obtain the so-called Carleman function solution of the Cauchy problem, cf. [37, 10.4], cf. [39, p 445].

Theorem 3.3. *Suppose that* (u,f) *is a solution of Maxwell's equations inside* X *, continuous up to the closure of* S *. Then*

$$\begin{pmatrix} u(x) \\ f(x) \end{pmatrix} = \lim_{N\to\infty} \int_S R_N^{\mathrm{i}}(x,y) \begin{pmatrix} in(f) \\ i\sigma(A^{\mathrm{i}})(v)t(u) \end{pmatrix} ds$$

for everyone . $x \in X \setminus \partial X$

Proof. We can assume by approximating X by smaller regions whose boundaries intersect the boundary of X only in , S that (u,f) is smooth in X . Combining Corollary 2.1 (Chapter 2) with Stokes' formula and Lemma 1.1, we easily conclude that

$$\begin{pmatrix} u(x) \\ f(x) \end{pmatrix} = \int_{\partial X} R_N^{\mathrm{i}}(x,y) \begin{pmatrix} in(f) \\ i\sigma(A^{\mathrm{i}})(v)t(u) \end{pmatrix} ds$$

for each fixed $x \in X \setminus \partial X$ and all $N = 0,1,...$. Let $N \to \infty$. Since the series (3.4.6) converges on $\partial B(0,R)$ uniformly over x on compact subsets of $B(0,R)$, it follows that the part of the boundary integral over $\partial X \setminus S$ tends to zero. This establishes the desired formulas.

Let (u,f) be the solution of the Cauchy problem (3.3.1). A simple calculation shows that

$$\int_S R_N^{\mathrm{i}}(x,y)\begin{pmatrix} in(f) \\ i\sigma(A^{\mathrm{i}})(v)t(u) \end{pmatrix} ds = G(u_0,f_0)(x) - \begin{pmatrix} U_N(x) \\ F_N(x) \end{pmatrix} \tag{3.5.2}$$

for $x \notin \overline{S}$, where

$$\begin{pmatrix} U_N(x) \\ F_N(x) \end{pmatrix} = \sum_{\nu=0}^{N} \int_S \Phi_\nu^{\mathrm{i}}(x,y)\begin{pmatrix} in(f) \\ i\sigma(A^{\mathrm{i}})(v)t(u) \end{pmatrix} ds .$$

Let us write $\varepsilon > 0$ for the distance between S and the origin. If $x \in B(0,\varepsilon)$, then the left-hand side of (3.5.2) tends to zero, since the series (3.4.6) converges uniformly to S . It follows that the series (i.e., the sequence of partial sums) $\{(U_N,F_N)\}$ converges to $G(u_0,f_0)$ uniformly together with its derivatives on compact subsets of the ball . $B(0,\varepsilon)$

The last observation and Theorem 1.1 combine to give some solvability conditions for the Cauchy problem.

Corollary 3.2. *If the series* $\{(U_N,F_N)\}$ *converges uniformly on compact subsets of the ball* $B(0,R)$ *, then the Cauchy problem (2.30) is solvable.*

Proof. Since the terms of the series $\{(U_N, F_N)\}$ are componentwise solutions of the Helmholtz equation, it follows from the Stiltjes-Vitali Theorem that its sum $(U,F) = \lim(U_N, F_N)$ satisfies in components the same equation in $B(0,R)$. Hence, (U,F) is the real analytic cross section of $F^i \oplus F^{i+1}$ on $B(0,R)$. Since (U,F) is effectively the same as $G(u_0, f_0)$ in the smaller ball $B(0,\varepsilon)$, the solvability of the Cauchy problem follows from Theorem 1.1. In many interesting cases, uniform convergence of the series $\{(U_N, F_N)\}$ is not only sufficient but also necessary for the solvability of (5.1), cf. [35; 2.9].

In this chapter, for an elliptic complex of first order differential operators on a smooth manifold, we define a system of equations which can be viewed as abstract Maxwell equations. The formal theory of this system turns out to be very similar to the theory of the classical Maxwell equations. We consider the Cauchy problem for the Maxwell equations in the theory of electrodynamics in a bounded region in Euclidean space.

LIST OF REFERENCES

1. Agmon S., Duglis A., Nirenberg L. Estimates of solutions of elliptic equations near the boundary. //M,1962.
2. Agranovich, M. S., and Vishik, M. I., "Elliptic problems with a parameter and parabolic problems of general form," UMN, 19:3 (117) (1964), 53-161.
3. L. A. Eisenberg and N. N. Tarkhanov, "Conditionally stable problems and Carleman formulas," //Sib. matem. zhurn. 31:6 (1990), 9-15.
4. Lavrentiev, M. M., Romanov, V. G., and S. P. Shishatsky, Uncorrected problems of mathematical physics and analysis, Nauka, M., 1980.
5. Maslennikova V.N. Differential equations in partial derivatives.// M., 1997.
6. Mergelyan S. N. , "Harmonic approximation and approximate solution of the Cauchy problem for the Laplace equation",// UMN, **11**:5(71) (1956), 3-26.
7. Petrovskii I.G. Lectures on equations with partial derivatives. //M.,1961.
8. Tarkhanov N.N. Loran's series for solutions of elliptic systems - //Novosibirsk: Nauka. Siberian Branch, 1991.-317 pp.
9. Tarkhanov N.N., Parametrix method in the theory of differential complexes - Novosibirsk:// Nauka. Sib. Branch, 1990 -248 p.
10. Wells R. Differential calculus on complex manifolds,//Moscow 1976

11. Fayazov K. S. , "On the Cauchy problem for a linear elliptic equation of second order with operator coefficients", Dokl. RAS, 336:4 (1994), 459-461

12. D. P. Fedchenko and A. A. Shlapunov, On the Cauchy problem for the multidimensional Cauchy-Riemann operator in the Lebesgue space L^2 in the domain, Math. lib. 199:11 (2008), 141-160.

13. Shabat B.V. Introduction to complex analysis, part H // M.: Nauka, 1985,219-223.

14. Shestakov I. V., On the Cauchy problem for Dolbo cohomologies,// Dissertation for the degree of Candidate of Physical and Mathematical Sciences, Krasnoyarsk, 2009.

15. Shlapunov A.A., On the Cauchy problem for the Laplace equation.Sibir. Mat. (1992) 3, **205-215**.

16. Shlapunov A.A., Tarkhanov N.N., To the Cauchy problem for holomorphic functions of Lebesgue class L^2 in the domain. Krasnoyarsk 1990.

17. Yarmukhamedov, Sh., "The Cauchy problem for the Laplace equation", Dok. AN SSR, 235:2 (1977), 281-283.

18. Sh. Yarmukhamedov, "Carleman function and Cauchy problem for the Laplace equation", Sib.matem. zhurnal, 45:3 (2004), 702-719.

19. Aizenberg L.A., Carleman's Formulas in Complex Analysis. //Theory and Applications, / Kluwer Academic Publishers, Dordrecht, 1993.

20. Bott R., On some recent interactions between mathematics and physics, //Canad. Math. Bull. 28 (2) (1985) 129-164.

21. Colton D., Kress R., Inverse Acoustic and Electromagnetic Scattering Theory, Springer-Verlag, Heidelberg, 1998.

22. Friedman B., Principles and Techniques of Applied Mathematics, //Dover Publications, Mineola, NY, 1990.

23. Harmuth H.F., Hussain M.G., Propagation of Electromagnetic Signals, //Word Scientific, Singapore, 1994.

24. Ikehata M., "Inverse conductivity problem in the infinite slab," Inverse Problems, 17:3 (2001), 437-454.

25. Ikehata M., "Two analytical formulae of the temperature inside a body by using partiallateral and initial data", Inverse Problems, 25:3 (2009), 035011.

26. Jackson J.D., Classical Electrodynamics, //John Wiley & Sons, Inc., 1975.

27. James Clerk Maxwell, A dynamical theory of the electromagnetic field, //Philos. Trans. R. Soc. Lond. 155 (1865) 459-512.

28. Kytmanov A. M., Nikitina T. N. Analogues of Carleman's formula for classical domains. Mat. notes, 1989, V.45, No.3, pp.87-93 (in Russian).

29. Ly I., Tarkhanov N., A variational approach to the Cauchy problem for nonlinear elliptic equations,// J. Inverse Ill-Posed Probl. Inverse Ill-Posed Probl. 17 (6) (2009) 595-610.

30. Makhmudov O., Niyozov I., Tarkhanov N, "The Cauchy problem of couple-stress elasticity",Complex Analysis and Dynamical Systems III, Contemp. Math., **455**, Amer. Math.Soc., Providence, RI, 2008, 297-310.

31. Monk P., Finite Element Methods for Maxwell's Equations,// Oxford University Press, Oxford, UK, 2003.

32. Schulze, B.-W., Boundary Value Problems and Singular Pseudo-Differential Operators, J. Wiley, Chichester, 1998.

33. Senior T., Volakis J., Approximate Boundary Conditions in Electromagnetics, //Institution of Electrical Engineers, London, UK, 1995.

34. Shestakov I. V. V. V., Shlapunov A. A., Negative Sobolev Spaces in the Cauchy Problem for the Cauchy- Riemann Operator, // Journal of SFU. Mathematics and Physics. 2009. №1. C. 17-30.

35. Shlapunov A., Green's integrals and their applications to elliptic systems, Scuola Normale Superiore, Pisa,// 1996, 162 pp., tesi di perfezionamento.

36. Stratton J.A., Electromagnetic Theory, //McGraw-Hill, New York, 1941.

37. Tarkhanov N. N. Complexes of Differential Operators, //Kluwer Academic Publishers, Dordrecht, NL, 1995.

38. Tarkhanov N.N., A criterion for solvability of the ill-posed Cauchy problem for elliptic systems, Dokl. Akad. Nauk SSSR 308(3) 531-534, 1989, Russian.

39. Tarkhanov, N. N. N. The Cauchy Problem for Solutions of Elliptic Equations, Math. Top., 7, Akademie Verlag, Berlin, 1995.
40. Tikhonov A.N., Samarskii A.A., Equations of Mathematical Physics, // Nauka, Moscow, 1972.
41. Yarmukhamedov Sh., Ishankulov T., Makhmudov O.I., The Cauchy problem for a system of equations in the theory of elasticity,// Siberian Math. J. 33 (1) (1992) 154-158.
42. Yarmukhamedov Sh., Yarmukhamedov I., "The Cauchy problem for the Helmholtz equation",Ill-Posed and Non-Classical Problems of Mathematical Physics and Analysis, VSP,Utrecht, 2003, 143-172.

43. Makhmudov K.O., Carleman formula for Helmholtz equation,// Abst. "Ukrainian math. Congress": Kiev,Ukraine, 2009.
44. Makhmudov K.O., "On the Cauchy problem for the Helmholtz equation", //International Trening-Seminars on Mathematics in conjunction with the joint mathematics meeting, between Samarkand State University and Malaysian mathematical sciences society , 3-5 june pp179-180 , 2011
45. Sattorov E.N., Makhmudov K.O., "About extension of a solution for the homogeneous system of Maxwell equations",// Uz. Math. J. **N3pp** 105-120, 2010
46. Makhmudov K. O., Makhmudov O. I., Tarkhanov N., "Equations of Maxwell type", //J. Math. Anal. Appl., 378:1 (2011), 64-75.(USA)

47. Makhmudov K.O., Ellipticity Maxwell's equations in the sense of A.Duglis-L.Nirenberg, // Scientific and practical seminar, "il-posed and non-classical problems of mathematical physics and analysis",5-6 julay Samarkand 2012.
48. Makhmudov K. O., Carleman formula in a cap type domain // "Actual questions Analysis" 22-23 aprel pp 141-146, Karshi-2016

49. Makhmudov K.O., Makhmudov O.I., Tarkhanov N., A Nonstandard
Cauchy problem for the heat equation, // J.Mat. Zametki 102 (2) (2017) 270-283.
50. Makhmudov K.O., A Nonclassical Cauchy problem for the generalized Maxwell equations // International conference devoted to the 60th anniversary of the S.L.Sobolev Institute of Mathematics. S.L. Sobolev Institute of Mathematics Novosibirsk, Russia August 14-19, 2017
51. Makhmudov K.O., Carleman formula for the Maxwell equation in a cap type domain, Jour. of Siberian Federal Uni.Math-s and Physics 12 (3)(2019) 317-322
52. Makhmudov K.O., Makhmudov O.I., Tarkhanov N., Nonstationary Maxwell equations, Complex Variables and Elliptic Equations,// 2020, Vol.65, No.6, 1029-1050.
53. Makhmudov K.O., Cauchy problem for the system of Maxwell's equations, //Scientific Journal of Mathematics-Informatics-Physics №1(125)(2021) 29-34

54.Makhmudov K.O., Dissertation work. "Cauchy problems for generalized system of Maxwell's equations". 2022-yil. Samarqand.

Printed by Books on Demand GmbH, Norderstedt / Germany